Matter and Meaning

Other volumes in this series:

Theology, Evolution and the Mind (ed. N. Spurway)

Creation and the Abrahamic Faiths (ed. N. Spurway)

Matter and Meaning:
Is Matter Sacred or Profane?

Edited by

Michael Fuller

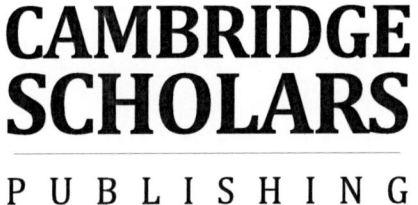

Matter and Meaning: Is Matter Sacred or Profane?, Edited by Michael Fuller

This book first published 2010

Cambridge Scholars Publishing

12 Back Chapman Street, Newcastle upon Tyne, NE6 2XX, UK

British Library Cataloguing in Publication Data
A catalogue record for this book is available from the British Library

ISBN (10): 1-4438-1907-7, ISBN (13): 978-1-4438-1907-7

TABLE OF CONTENTS

Preface vii
Rt Revd James Jones, Bishop of Liverpool

Chapter One 1
Introduction: Matter and Meaning
Michael Fuller

PART I: SCIENTIFIC PERSPECTIVES

Chapter Two 9
Why Matter? A Scientific Perspective
Ruth Gregory

Chapter Three 21
Re-Creation: A Possible Interpretation of Quantum Indeterminism
M. Basil Altaie

PART II: HISTORICAL PERSPECTIVES

Chapter Four 39
Theology and Matter Theory in the Early Modern Period
Peter Harrison

Chapter Five 57
Theology and the Meaning of Matter: A Response to Peter Harrison
John Henry

Chapter Six 67
Models and Symbols in the Understanding of Matter
Colin A. Russell

Chapter Seven 83
A Response to Colin Russell's *Models and Symbols
in the Understanding of Matter*
Michael Poole

Chapter Eight.. 87
Matter: An Islamic Perspective
M. Basil Altaie

PART III: THEOLOGICAL PERSPECTIVES

Chapter Nine... 103
The Triune God and the Triad of Matter
Niels Henrik Gregersen

Chapter Ten 119
A Response to Niels Henrik Gregersen's *God, Matter and Information*
Kenneth Wilson

Chapter Eleven 123
Divine Grace and the Created Order in the History of Catholic Theology
Hilary C. Martin

Chapter Twelve 131
God and the Matter Delusion: The Denial of Matter in the Teachings
and Practice of Christian Science
Daniel Scott

Chapter Thirteen... 141
On the Capacity of Sound Waves, Painted Canvas and Printed Page
to Carry Meaning – And the Place of the Arts in a New-style Natural
Theology
Peter Barrett

Contributors... 153

Index.. 157

PREFACE

THE RIGHT REVERED JAMES JONES, BISHOP OF LIVERPOOL

When the Science and Religion Forum met in conference at Liverpool Hope University I was honoured to be invited to address them after dinner on their final evening. I was deeply impressed by the scope and the expertise of the symposium, not just of those giving papers but of the members themselves. At a time when siren voices in the media want to polarise science and religion, it is immensely hopeful that we are served by scholars who are able not only to bridge that divide but also to show by example and research that there is a consonance and a correspondence between these two scholarly disciplines.

The symposium took place in Liverpool during its year as European Capital of Culture. Science and religion constitute two pillars of European civilisation; and this was illustrated shortly before the symposium took place. On the same evening in the City Professor Richard Dawkins packed the auditorium at the Philharmonic Hall, while a stone's throw down Hope Street an equally large audience had been enthralled by the world premiere of Sir John Tavener's 'Requiem' in the Metropolitan Cathedral. Whatever polarities were represented by these two events it was good then to welcome to the City the Science and Religion Forum for their symposium on 'Matter and Meaning: Is matter sacred or profane?'

I was particularly interested to engage with the Symposium on this subject because of my increasing conviction that Christianity has vitally important insights to share with the world in the face of the current ecological crisis. One of the reasons that the Christian faith has been slow off the starting blocks in raising awareness of our environmental responsibilities is because we have given priority to the spiritual over the material. Indeed, some would argue that allowing a division between the two, and failing to recognise that the material and spiritual are indivisible, marked the beginning of the retreat from engaging with the full reality of God's world.

The Lord's Prayer is a petition for God's will to be done on earth as it is done in heaven. This is a prayer for the earthing of heaven. I like the version from the Book of Common Prayer, which has us asking for God's will to be done 'in earth'. The preposition 'in' gives a depth to the meaning, and signifies that the spiritual work of God's Kingdom, or bringing heaven down to earth, is fundamentally and essentially material.

There is a spiritual and material continuity from Genesis to the Gospels. At the outset God declares his world in all its dimensions to be 'good' and 'very good'. The Resurrection of Jesus – his whole self, not just his soul – is the divine affirmation that God's purposes include the material as well as the spiritual. The Resurrection of the body in the Gospels shows that matter matters to God, which is the creed to be found in Genesis.

If it were only the soul that continued after death there might be some excuse for relegating the physical to a place of less importance. One of the conundrums of the last century is why the evangelical tradition, which was so emphatic about the bodily dimension to the Resurrection of Jesus, was so suspicious of the social Gospel – and why the Liberal tradition, which sat so loose to the Resurrection of the body, was so strong about the material dimensions of the Kingdom. Fortunately, these separations are less fixed today; but they have dogged the mission of God in the not too distant past, and have frustrated those who long to see Science and Religion marching in step with each other.

The sacredness of matter has major repercussions on how religious believers engage with politics, sexuality and the arts as well as with the sciences. It also has an impact on attitudes to the environment. In this country, in America and in Africa, I've heard Christians suggest that we need not concern ourselves with what is happening to the earth because, according to the Bible, it will end up in a ball of flames. Some have even gone on to suggest that we might as well milk the earth for all it is worth while we have time!

These attitudes to the earth make the task of the environmentalist and the scientist more challenging, especially if they are people of faith. It suggests that within some quarters of the community of faith the priority lies in preparing people for what is to come beyond the consummation. It suggests that exploring, analysing, understanding and caring for the earth are secondary activities. Little wonder that scientists who are believers might find themselves undervalued within the community of faith.

What we need is a theology of the earth that recognises the Biblical vision of a world that is originally good, created by God and sustained by him; a world that has come into being through and for Christ, and that will be renewed and transformed for the glory of God. Such a world is the scientist's laboratory, and the believer's temple.

Recently, re-reading C.S. Lewis' 'The Last Battle', I was struck by the depiction of devastation as old Narnia came to an end. It resembled a scene of deforestation.

> They went to and fro tearing up the trees by the roots and crunching them up as if they were sticks of rhubarb. Minute by minute the forests disappeared. The whole country became bare. The grass died. Soon Tirian found that he was looking at a world of bare rock and earth. You could hardly believe that anything had ever lived there.

The children are understandably distressed by all the destruction and turn to Lord Digory, who dismisses all that they had witnessed as 'only a shadow or a copy of the real Narnia which has always been here and always will be here'. He differentiates the new Narnia from the old 'as different as a real thing is from a shadow or as waking life is from a dream'. But what leapt off the page was the basis for Lord Digory's explanation: '.... he added under his breath, "it's all in Plato, all in Plato: bless me, what do they teach them at these schools!"'

It made me wonder how much Platonic thought informed our present world view, and how much we need to affirm that the earth is indeed 'the real thing' and neither a dream nor a passing shadow. The earth is the Lord's – and the place in which the believer prays for God's will to be done, and in which the scientist explores the mysteries of the universe.

I ended my address to the Symposium with the beautiful imagery of Gerard Manley Hopkins' poem 'God's Grandeur'.

> The world is charged with the grandeur of God.
>   It will flame out, like shining from shook foil;
>   It gathers to a greatness, like the ooze of oil
> Crushed. Why do men then now not reck his rod?
> Generations have trod, have trod, have trod;
>   And all is seared with trade; bleared, smeared with toil;
>   And wears man's smudge and shares man's smell: the soil
> Is bare now, nor can foot feel, being shod.
>
> And for all this, nature is never spent;
>   There lives the dearest freshness deep down things;
> And though the last lights off the black West went
>   Oh, morning, at the brown brink eastward, springs –
> Because the Holy Ghost over the bent
>   World broods with warm breast and with ah! bright wings.

It was good to be in the company of those whose minds can span the disciplines of science and religion, and whose imaginations can behold the grandeur of God flowing out of the material world 'like shining from shook foil'.

CHAPTER ONE

INTRODUCTION:
MATTER AND MEANING

MICHAEL FULLER,
THEOLOGICAL INSTITUTE OF THE SCOTTISH
EPISCOPAL CHURCH

The title of this symposium may be seen to be a very conscious drawing-together of the worlds of science and theology. Scientists explore matter: they investigate the stuff of which the physical world is made, and the ways in which that stuff interacts and combines to form the things which we observe around us. Theologians explore issues of meaning, and of purpose. Put like this, we may initially suppose that our title confronts us with a classic case of the 'independence' model of science and theology, of Gould's 'non-overlapping magisteria' (cf. Barbour 1998, p. 84 ff; Gould 2001). But is this, in fact, a topic on which fruitful dialogue between these two areas of human endeavour can occur? The contributors to this volume certainly think so. Indeed, it would be odd for a theologian, at least, to think otherwise. One need not advocate a return to a Paleyesque natural theology to believe that, if our cosmos is in some sense the creation of God, then the matter from which that cosmos is constituted may have a story to tell other than simply that of what it is comprised.

But this is, of course, to beg the initial question: in what sense (if any) can matter be said to *mean* anything at all? Here, perhaps we may usefully distinguish between three approaches.

The first approach, which has characterised much Christian thinking down the centuries, is to see the material world as shot through with meaning. An exquisite example of this is afforded by Gerard Manley Hopkins' poem 'God's Grandeur', quoted in the preface to this volume, with its assertion that 'The world is charged with the grandeur of God'. Whilst the Bible is by no means univocal concerning the importance and value of material things, the psalmist's declaration that 'The heavens are telling the glory of God; and the firmament proclaims his handiwork' (Psalm 19:1), may be said to exemplify such a view. The material world has the power to inspire people to reflect upon the meanings it may have; and whilst evolutionary theorists have not been slow to speculate on why humans should possess such a capacity for feeling inspired (cf. Boyer 2002, Barrett 2009), such speculations scarcely negate that capacity.

The second approach would be to say that matter is inherently meaningless. This is part of a world-view characterised by Keith Ward as materialism: 'Materialism says that the only things that exist are material things in space. There is no purpose or meaning in the universe. Scientific principles are the only proper forms of explanation' (Ward 1996, p. 99). Ward associates this viewpoint with zoologist and scientific populariser Richard Dawkins, who writes that 'a body is really a machine blindly programmed by its selfish genes' (Dawkins 1989, p. 146). Genes, of course, are contained within the DNA found in living organisms; and Dawkins powerfully advocates the view that 'DNA just is' – that it, in common with other physical material, is simply a phenomenon found in a universe which he declares to consist of 'just electrons and selfish genes' (Dawkins 1996, p. 155). According to this viewpoint, then, matter 'just is': it is meaningless: it is simply something which is there to be studied (and science is the appropriate systematic method for such study).

The third approach would be somewhere in between these two. We may not wish simply to latch on to the insights of a particular set of texts affirmed as normative within a religious tradition, being conscious that such texts necessarily reflect the outlook of the age in which they were produced; but we may also feel unsatisfied with a Dawkinsian dismissal of meaning altogether. Rather, we might affirm that whilst the data of science in and of themselves may be 'meaningless', by placing them within the context of a narrative framework meaning may be made of them. This is, in fact, what Dawkins does: but the narrative within which he chooses to understand the raw data obtained by scientists is one which presumes a metaphysical materialist outlook, with its explicit denial of meaning. (For a discussion of narratives, including the Dawkinsian scientistic narrative, as conveyors of meaning, see Smith 2003, esp. chapter 4.) There can be no doubt that Dawkins tells his tale with skill and verve. But other narratives, other stories, might yet allow for understandings which would see matter as meaningful, without imposing any distortions on the data of science or their rigorous interpretation. The choice of which story to tell lies with the storyteller; and the choice of which narrative is to be believed lies with the hearer of the tale.

The relationship of human beings with matter, as investigators of it, as re-presenters of it, as interpreters of it, and as composers of narratives concerning it, is a complex one (cf. Polanyi 1958, Polkinghorne 1991). In a book titled 'Atoms and Icons', I noted that 'It is atoms which compose the face on an icon, and it is icons which are used to represent the invisible world of the atom' (Fuller 1995, p. 146). The symbolic creation of the icon-painter is clearly intended to be meaningful; yet it is a material object, albeit a carefully-crafted one. Conversely, the very description of material objects – particularly objects like atoms, the complexities of which may not be observed directly – may of necessity require symbolic representation; and symbols inevitably bring with them an epistemic flexibility which opens up the possibility of their conveying a variety of meanings, not solely those intended by the originator of the symbol.

It is here that we may see a possible ground for the interaction of scientists and theologians which is ripe with possibility. If we set to one side those narratives that preclude the involvement of one of those dialogical partners, be it the historical unpacking of a scriptural narrative in the absence of scientific insight or the materialist paradigm refusing to countenance meaning in a universe consisting solely of 'electrons and selfish genes', then many fruitful opportunities for conversation open up. And in fact, as contributors to this volume note, such dialogue would appear historically to have been the norm rather than the exception.

*   *   *   *

The papers delivered at the Science and Religion Forum's 2008 conference at Liverpool Hope University have been gathered together in the present volume into three parts. Part One offers accounts of our current scientific understanding of what matter is. Ruth Gregory describes the remarkable developments in twentieth century science which have built up our current thinking regarding how matter at its most basic level is constituted, and how it behaves. She introduces the mathematical world underpinning and shaping our understanding of contemporary science, through a consideration of the ideas of those who were its founding fathers: Planck, Einstein and Dirac, amongst others. Gregory's paper also demonstrates something of the 'unreasonable effectiveness of mathematics', as Eugene Wigner put it, in enabling us to describe and account for the most remote material systems to which we can have access – those at the atomic and subatomic levels. Basil Ataie also offers some reflections on this historical account, and then explores some of the ways in which quantum measurement has been understood. He also proposes a novel understanding of quantum indeterminism, in terms of the physical properties of micro-systems being in a constant state of re-creation.

Part Two offers some historical perspectives on the ways in which we currently understand matter. Peter Harrison extends the historical canvas under consideration further back through history, to consider formative thinkers prior to the twentieth century. His paper focuses in particular on those classical thinkers who saw a dualism between what we would now term the world of matter and the world of ideas, or the world of spirit: thinking which has cast a highly significant shadow over subsequent centuries. Against this backdrop, Harrison suggests ways in which the corpuscular and mechanical views of nature, which gave rise to science in the early modern period, have tended to drain 'meaning' from the natural world. Given the explanatory power of the theories spawned by this understanding of nature, it is clear that the presumption of the meaninglessness of matter is now firmly engrained in the Western psyche, as a metaphysical assumption that can often be unquestioningly accepted by those who turn to the sciences for explanations of natural phenomena. John Henry, in a response to this paper, demonstrates the irony that these changes happened principally as a consequence of the attitude of the Church, rather than as a reaction against the Church's teachings.

Colin Russell's paper offers some further historical insights into the ways in which metaphors and models have been used and developed throughout history to generate explanations of the understandings of matter current at particular times. In a valuable 'case study' illustrating this, he draws attention to the present global warming crisis; and he suggests that this is the point as which theological thinking might intervene. Thinking of matter not as something objective, which is there to be exploited, but rather as something subjective, to which we can relate, assists us in facing up to incipient climate change; and, Russell adds, it may enable us also to re-introduce the idea of sacredness into our understanding of matter. Russell's clear inference is that this is an understanding that serves us rather better than ones which have prevailed during the post-Enlightenment centuries: centuries which, famously, have seen a desacralisation of nature.

Michael Poole responds directly to a number of points made in Russell's paper, commenting on and drawing out some of the ideas found there. Concluding this section, Basil Altaie offers a valuable view from an Islamic perspective, reminding us that scientific thinking flourished within this tradition for centuries before the development of Western science as we know it today. Altaie also points out some of the ways in which insights from Islamic thought can feed directly into 'hot' topics in today's dialogue of science and theology.

In Part Three, contributors present some theological perspectives which explore the interpretation of matter as meaningful. Niels Gregersen's paper reflects on contemporary understandings of matter, and draws on information theory and on logos-theology in noting some fascinating parallels between theological and scientific enquiry – in particular, that the natural 'triad' of mass, energy and information suggests constructive parallels with traditional Christian Trinitarian thinking about God, as Father, Son and Holy Spirit. Kenneth Wilson, responding to Gregersen's paper, draws on Wittgenstein and others in unpacking further some of the points being made. Hilary Martin explores a Roman Catholic understanding of the way in which the relationship of nature and grace may provide a perspective on the created order, which in turn promotes a greater integration of divine grace and earthly reality. Daniel Scott explores the radically anti-materialist thinking of Mary Baker Eddy, and examines the ways in which this has found expression in the outlook of the Christian Science movement. Finally, Peter Barrett draws on the writings of John Polkinghorne, Anthony Monti and others in developing a 'New Natural Theology' which includes the insights of the arts.

Many people in the twenty-first century West doubtless continue to think of matter as the inert 'stuff' of the universe, no longer 'sacred' in a way in which it might have been understood in the past, and belonging to a category of existent for which the term 'profane' would be equally inappropriate. The historical studies presented here trace something of the story by which this understanding of matter has arisen. But the contributors to this volume suggest that this is not the end of the story; and Russell's writing of a *resacralisation* of matter, together with Gregersen's invitation to think of modern relativistic and quantum theories in terms of their *dematerialisation* of matter, suggest that other understandings of the 'stuff' of our universe are possible. Perhaps the materialist metaphysic which has become habitual for many in the twenty-first century West is better exchanged for a rather different view. This view would see matter as imbued with the sacred (which, for believers in God, will mean that it has some capabilities for revealing to us something about the divine nature – and which for everyone, believers in God or not, will serve as a reminder of the ethical responsibilities which we bear when we manipulate it). And this view would see matter as something of which the absolute nature eludes us more and more the deeper it is studied – and would hence urge humility on those who seek to understand it further.

Bibliography

Barbour, I. G. 1998. *Religion and Science: Historical and Contemporary Perspectives*. London: SCM Press.

Barrett, J. 2009. 'Cognitive Science, Religion and Theology', in Jeffrey Schloss and Michael J. Murray (eds.), *The Believing Primate: Scientific, Philosophical and Theological Reflections on the Origin of Religion*, pp. 76-99. Oxford: Oxford University Press.

Boyer, P. 2002. *Religion Explained*. London: Vintage

Dawkins, R. 1989. *The Selfish Gene* (new edition). Oxford: Oxford University Press.

Dawkins, R. 1996. *River out of Eden*. London: Phoenix.

Fuller, M. 1995. *Atoms and Icons: A Discussion of the Relationships between Science and Theology*. London: Mowbray.

Gould, S. J. 2001. *Rocks of Ages: Science and Religion in the Fullness of Life*. London: Jonathan Cape.

Polanyi, M. 1958. *Personal Knowledge*. London: Routledge and Kegan Paul.

Polkinghorne, J. 1991. *Reason and Reality: The Relationship between Science and Theology*. London: SPCK.

Smith, C. 2003. *Moral, Believing Animals: Human Personhood and Culture*. Oxford: Oxford University Press.

Ward, K. 1996. *God, Chance and Necessity*. Oxford: Oneworld Publications.

PART I:

SCIENTIFIC PERSPECTIVES ON MATTER

CHAPTER TWO

WHY MATTER?
A SCIENTIFIC PERSPECTIVE

RUTH GREGORY,
UNIVERSITY OF DURHAM

The recent switching on of the Large Hadron Collider (LHC) at CERN in Geneva makes my topic particularly timely: the physics of matter, or, for fans of *Angels and Demons*, antimatter. The experiments in Geneva are aimed both at verifying our theories of how matter works, and also at pushing forward the frontiers of our understanding and hopefully uncovering something new and, maybe, something unexpected. Although I will not cover some of the more technical aspects of our theory of particle physics, the *Standard Model* as it is known, I will discuss the underlying physical principles that lead to it. Like many scientists, I concern myself largely with the question of *how* the universe works, rather than *why*. Nevertheless, our struggle to understand the 'how' often leads us to some very interesting 'why's!

The foundation for understanding what we mean by matter is Quantum Mechanics. Even those who developed it considered this theory paradoxical: Niels Bohr said, 'Anyone who is not shocked by Quantum Mechanics has not understood it'. It touches almost every aspect of our modern daily life. Imagine no TV, computers, iPods, or Nintendo; nor any medical diagnostics and, some would argue, no free will. Quantum Mechanics is inherently paradoxical from a non-scientific standpoint. It explains how the atom splits, and yet how it is stable; it gives us uncertainty, yet also makes predictions for experiments. So how did we get to Quantum Mechanics?

Democritus, the laughing philosopher, believed that matter could not be subdivided *ad infinitum*, but that there had to be a 'smallest unit' – the atom. His basic philosophy was that the continuum (as we now call it) did not exist. To a large extent, this must have been a matter of faith, as there was no empirical evidence at the time for either finite or infinitesimal subdivisions. It is remarkable that, in the absence of experiment, he could have intuited the essence of matter. Although we often focus on its shortcomings, the scientific world view of the Greeks was breathtakingly successful in its reach, and indeed, based as it was on aesthetics, is very analogous to fundamental high energy physics today

The concept of an atom was formalized by John Dalton around 200 years ago. He laid down a set of hypotheses governing the nature of matter, and how it interacted chemically. At that time, chemical reactions were the only scientific basis for experiment. His definitions are recognizable to any student of chemistry these days:

> Matter is made up of **atoms**, which are **indivisible**.
> All atoms of a given element are **identical**, but different elements have different atoms.
> Atoms cannot be created or destroyed, only rearranged in chemical reactions.

Later in the nineteenth century, Dmitri Mendeleev classified the elements by the nature of their chemical interactions. The great significance of this work was that it showed patterns in the atoms of nature. We now understand the periodic table in terms of the structure of electron orbits around the atomic nucleus; but this understanding took a great deal of time to dawn.

The first step on the road to quantum mechanics started with the theory of heat, developed towards the end of the nineteenth century. James Clerk Maxwell and Ludwig Boltzmann developed the theory of *Statistical Mechanics*, whereby heat (and all other thermodynamical quantities) was described in terms of bulk motion of the constituent particles. Thus heat was simply atomic or molecular motion: the hotter the gas, for example, the more energy the molecules had. The statistical aspect comes because there are such large numbers of molecules, and we do not need to know the detail of their behaviour, just the overall net effect.

However, a problem arises, in that according to statistical mechanics, all the energy is divided democratically between the available states; and for a hot radiating body, this means that light of all wavelengths should be radiated. But there are infinitely many wavelengths! This means that the amount of energy radiated in visible light should be zero – but we know this isn't true. In fact, a sharp cut-off in the ultraviolet is seen in the spectrum from a hot body, and no light is radiated at very short wavelengths. This paradox (or rather, the failure of otherwise good scientific theories to explain a known phenomenon) became known as the *ultraviolet catastrophe.*

Meanwhile, J. J. Thomson had discovered the electron, so physicists now knew that in fact atoms were not indivisible. Even worse – when he repeated his experiment to try to observe positively charged ions, which should be the parts of the atom identifiable with the chemical element, he found that they were not in fact identical. Rather, ions of the same element could have different masses (Thomson called these *isotopes*). The simplest example is hydrogen, which has three isotopes: hydrogen, deuterium and tritium.

Finally, the discovery of radioactivity showed that atoms could transmute into other atoms. It seemed that all of Dalton's hypotheses had fallen: atoms were not identical, indivisible or fundamental. So, what was 'matter' now?

Once it was realized that atoms had structure, the next aim was to determine what that structure was. In 1908, Rutherford was firing alpha particles (these are helium nuclei: small positively charged objects emitted during radioactive decay) at gold film to prove his 'plum pudding' atomic model. He believed that atoms were structured as a globule of positive material, with negatively charged electrons embedded in them like plums in a pudding. He expected to see the alpha particles deflected by the positive pudding, emerging at a variety of different angles. Instead, he found that most of the alpha particles went straight through the gold film, with just a few bouncing back. This was only possible if the positive part of the atom was tiny and concentrated, and Rutherford deduced that the structure of the atom was more like a mini solar system: a tiny positively-charged nucleus surrounded by electrons orbiting around this centre. Rutherford could estimate the size of the nucleus from the scattering, and found it was less than one millionth of a millionth of a centimeter – roughly the equivalent of a pinhead in the centre of a football field. This was a complete shock: the plum pudding had been replaced by empty space! This immediately led to another problem: electrons in orbit around a positive charge will radiate by Maxwell's theory of electromagnetism – so how was the atom stable?

Fortunately, as is often the case when ideas and understanding are developing so rapidly, the tools which could supply an answer were already on the shelf. Six years previously, Max Planck had provided a solution to the ultraviolet catastrophe. He noticed that if you assumed that energy came in lumps, rather than being continuous, and moreover if you assumed that those units of energy increased as the wavelength of light decreased, the black body spectrum could be explained. He proposed the following equation:

$$E = h / \lambda$$

where E is the energy contained in the little lump of light (known as a *photon*), λ is its wavelength, and h is a constant of proportionality, now known as Planck's constant.

It is worth just taking a moment to reflect on Planck's equation, as this is absolutely central to the development of quantum mechanics. His claim was that light at a particular wavelength (or frequency, the two being related through the speed of light) could only contain multiples of a fundamental unit or *quantum*. Since light can only be emitted with *at least* this quantum of energy, which increases with decreasing wavelength, a radiating body will not have enough energy to radiate at very short wavelength; hence the cut-off of the black body spectrum in the ultraviolet.

Planck at first did not believe his idea was fundamental, corresponding to some underlying reality, since what it actually says is that light comes in units – it is a *particle*. But everyone knew light was a wave: James Clerk Maxwell had demonstrated that with his theory of electromagnetism forty years earlier. Indeed, some two centuries earlier Huygens had correctly explained refraction and polarization in terms of wavefronts of light. However, another physicist at the time was prepared to believe Planck.

1905 was an amazing year for physics: it was the year that Albert Einstein emerged onto the scene. Even the least of Einstein's papers that year would be a feather in the cap of most physicists today. Yet, as well as ultimately overturning our concepts of space and time, and explaining Brownian motion (the random motion of larger particles in a fluid) and other statistical properties of matter, Einstein set a revolution in motion. He took Planck seriously, and by assuming the quantization of light he was able to explain the photoelectric effect. In the photoelectric effect, light shining on certain metals releases electrons of the same velocity, independent of the intensity of the light: increasing the light intensity leads to more electrons, not more energetic electrons. At the time, this was a puzzle – after all, if you hit something harder it tends to move faster, so why were the electrons all emerging with the same speed? Einstein explained this by supposing that the light consisted of Planck's light quanta – the photons – and that only a photon of the right wavelength could knock an electron out of the metal. Adding more photons would not make the electron move any faster: it could only release more electrons.

At first the scientific community was skeptical, but in 1913 Niels Bohr realized that this quantization could explain both atomic structure and stability. Bohr suggested that the energy of orbiting electrons was quantized, and that the electron could jump between orbits if given the right 'kick' from a photon. In particular, this meant there was a lowest orbit, which was stable, and it also explained *and predicted* characteristic lines in the spectrum of the hydrogen atom. It was not a perfect description, since Bohr assumed that orbits were circular, and this does not work so well for electrons in higher orbits, or for elements with a higher atomic number than Hydrogen. However, it did capture the idea of the quantum atom, rather like Ptolemy's universe captures the idea of the solar system, and this model is still taught to students today to illustrate the application of quantum mechanics to the atom.

The penultimate piece of the quantum puzzle was put in place by Louis de Broglie in 1924, when he proposed wave-particle duality. Just as the ideas of Planck and Einstein show that light can have a particle-like nature, de Broglie suggested that particles like electrons could have a wave-like nature. He inverted Planck's equation to say that a *particle* would have a fundamental *wavelength*, the de Broglie wavelength as it is now known. Again, the amazing feat here was not that de Broglie reversed the order of an equation, but that he reversed the *interpretation*, and drew a far-reaching and revolutionary conclusion from this hypothesis. Not only could light, a wave, behave like a particle, but also particles could behave like waves. Using Einstein's relation between energy and momentum for a photon, the de Broglie wavelength is defined as

$$\lambda = h / p.$$

Here, h is Planck's constant as before, and p is the momentum of the particle. (Recall that momentum, which is conserved in collisions, is the combination of mass and velocity: $p = mv$). Why had nobody noticed this extraordinary conclusion before? The reason we are not aware of our wave-like nature is that the wavelength is so small. Planck's constant is an extremely tiny number, and so the de Broglie wavelength of a human would be one million billion billion billionth of a centimeter (10^{-33} cm)! However, the de Broglie wavelength of an electron is the size of an atomic radius, so the size of the atom emerges naturally from this description. Moreover, de Broglie could explain why Bohr's electrons orbited the way they did. They were simply standing waves, with each orbit a given 'multiple' of the electron's wavelength.

At this stage, everything was conceptually in place; the fundamental nature of matter was more or less understood, at least in essence. However, science is not about general descriptions or conceptual understanding, essential though these things are. Science is about making predictions, quantifying results, and measuring the outcomes of an experiment. Science is about the construction and verification of a mathematical model of nature: in other words, a theory.

In 1926, Erwin Schrödinger presented his 'wave equation'. This was a theory of how quantum particles behave. It incorporated de Broglie's idea of waves by replacing the electron by a *wave function*, a number at every point in space and in time. In a similar way to Maxwell's equations for the propagation of light, Schrödinger gave an equation for the way this wave function evolved. However, this wave function was something very different from the usual expressions used in physics. Not only was it a complex number, but also it did not directly correspond to anything concrete, such as the size of an electric field. Instead, Schrödinger's wave function encodes the *probability* of finding the electron at a particular point in space (and time). It does not represent the electron as we might visualize it in our minds, but more the electron in its full generality. Thus, Schrödinger had made the final conceptual leap from the classical predictability of Newton and Maxwell, and introduced a quantum world: a place of probability, uncertainty and chance.

Complex numbers are a tool by which scientists generalize the real, or ordinary, numbers. They have a 'phase' as well as a 'size'. They not only allow you to take roots of negative numbers, but they are also very powerful mathematically. From another perspective, they contain more information than you can actually 'see'. There is a reason that counting or number systems have been with us for millennia, but complex numbers for only a few centuries: our perception of the world is real. In part because of the extra hidden information in complex numbers, but also because we have replaced a concrete thing with a probability, we are led to a property known as *uncertainty*. If we observe a particle, we interact with it, which changes it. We can never know exactly where it is without completely destroying our knowledge of where it is going. In other words, we no longer know where we are, where we are going, how much we've got and when we are going to get there – at least, not all at the same time!

It is worth pausing for a moment to reflect on where our journey into the quantum world has led us. We have, step by logical step, been forced to blur the distinction between forces, like electromagnetism, and the objects those forces act on, like electrons. In the 19th century, these were either waves or particles, and real. Now, we see they are each both wave-like and particle-like, and possibly complex. We find that our description of nature includes not only things that we do not see, but also things that we can *never* see.

Einstein found this deeply disturbing, and felt that nature could not be fundamentally indeterministic. 'God does not play at dice', was his famous complaint: 'Stop telling God what to do', was Bohr's laconic response! In many ways, Quantum Mechanics was Einstein's unruly child. Although he had fathered the theory, it had become something with which he could not reconcile himself. Yet we now know that this understanding of nature is absolutely right.

However, the Schrödinger equation was not the final word. As Schrödinger was well aware, it was not directly compatible with Einstein's Special Relativity, a well-verified theory. Schrödinger in fact had initially derived his wave equation in a more conventional form. To understand how he did this, we can think of the equation as a sum of energies, which reads much as a classical Newtonian relation: the total energy is a sum of kinetic and potential parts. However, Newton thought time was absolute, and certainly separate from space. From Einstein, we know that space and time have to be on the same footing. This means that every time we see a length, we need a time to balance it. For the energy relation, it actually means it is a sum of squares, like Pythagoras:

$$E^2 = m_o{}^2 c^4 + p^2 c^2$$

Here, we see the famous $E = mc^2$ relation of Einstein, the energy contained in matter. We also see the kinetic energy, contained in p, the momentum. Translating this into our wave function, we relate E to a rate of change in time, and p to a gradient in space. This gives a relativistic equation, the one Schrödinger originally obtained, with time and space appearing on the same footing. Schrödinger abandoned this original form because the square of the energy appeared. This meant that the energy was given by a square, and hence there could be negative as well as positive energy solutions; but how could energy be negative?

Schrödinger believed that a negative energy solution to his wave equation was a disaster. The equation would then predict that negative energy particles would be produced, which would actually lower the energy further, so more would be produced, leading to a runaway instability of the vacuum. He could find no satisfactory way to avoid this negative energy solution, and so he abandoned the relativistic equation, taking an approximation for low kinetic energy. It is not the first time in physics that a rough and ready working model has turned out to be more valuable than the 'Rolls Royce' version. The Schrödinger equation is used in most modern atomic physics and nanotechnology.

A theoretical particle physicist, however, seeks to describe nature as accurately as possible; and since nature is relativistic, it was essential to understand what happened to the negative root. Paul Dirac, a Cambridge mathematical physicist, believed that if we could take the square root of the equation directly, then the problem of negativity would go away. The trouble was, relativity implied that space and time should be on the same footing; but there are three dimensions of space and only one of time. Dirac had a moment of inspiration when he realized that by mopping up these single gradients by an array, or matrix, of other numbers, he could take a meaningful relativistic square root. He wrote down his 'gamma matrices', deriving the relations they had to satisfy, and he then had the insight that indeed they could satisfy those relations if the wave function became a more complicated expression known as a *spinor*.

While working through his theory, Dirac found that he had not removed the negative energy. Rather, he had simply given it a new place to hide in the extra information contained in the wave function. By this point, however, he was sure he was on the right track, and sought an alternative explanation. He theorized that there were negative energy solutions, but that they would in fact be full of electrons. Electrons were known to obey the Pauli exclusion principle, which states that no two electrons can occupy the same state. (This principle ultimately has a neat explanation from the Dirac equation.) It was therefore quite possible for these negative states all to be filled. The true vacuum was then a state in which all the negative energy solutions were populated. Dirac interpreted a hole in the 'sea' of negative energy states as a positive-energy, positively-charged particle, which would form if an electron were kicked out of a negative state. This new particle would therefore have the *same* mass but *opposite* charge as the electron.

Dirac thus predicted a new particle. Four years later this was observed by Carl Anderson, who christened it the 'positron', thus heralding a new era of particle prediction and discovery in high energy particle physics. We now have the idea of an anti-particle to every particle, which is essentially 'the same but opposite' to the particle. An anti-particle has all the same charges as the particle, but with the opposite sign. The only thing a particle and anti-particle do not possess in opposite degree is their energy, which is positive. What this means is that when a particle meets its anti-particle, there is nothing to stop them from unravelling each other, i.e. they annihilate.

This is what is used in the positron emission tomography (PET) scanner; the precisely collinear photons produced from such annihilations give a precise location of the decay, which allows for extremely accurate imaging. The energy released in a single decay is tiny, which is why it is a safe diagnostic tool, but if we had even a small amount (by everyday standards) of antimatter, the corresponding release of energy would be huge. For example, one kilogram of matter and antimatter would release 270 *billion* kWh of energy!

The Dirac equation underlies the basic description of most 'matter'. But what do we mean by 'matter'? From a particle physics point of view, matter is something which we can describe by the Dirac (or other appropriate) equation, where we identify the particle through its charges: most particles carry not just electric charge, but also other hidden charges which we do not see at our macroscopic level. The antiparticle then has the opposite charges. This is the particle physics picture, but from a more mundane point of view we imagine matter to be the stuff we are made of – something with mass, and possibly with some electric charge. Why therefore do we not see other charges, why is matter massive, and why do we not see anti-matter? These are the issues I now briefly explore in the closing part of this chapter.

Most familiar (and less familiar) particles have a property known as spin, which can be thought of as the particle spinning on its axis rather like the earth or the sun. With one rather high profile exception, all known particles have spin, that exception being the Higgs boson, which has no spin at all. The Large Hadron Collider (LHC) has been built primarily to find this final piece of the Standard Model of particle physics, as well as looking for clues beyond the Standard Model. What makes the Standard Model so elegant is that is encodes the known quantum interactions of nature in an economical description, with a relatively small number of fields. The model has hidden symmetries relating different particles through this as yet unobserved Higgs scalar. Observing the Higgs boson would tell us that we have the right picture of nature and also confirm our theories of how particles get their masses, and how these masses relate to each other. But of course that is not the whole story. We want to push our theories beyond this Standard Model, because like Einstein we want to incorporate gravity into particle physics. This very interesting story is too long to review here, so I would like to finish off with a few remarks about the Universe, cosmology, and what we do not know – which is how matter got here in the first place.

Let me first explain in a nutshell the standard cosmological model. This is an astonishing achievement of 20th century physics, using rather broad and basic theories and models to achieve a staggeringly successful description of the cosmos. Once again, we start with Einstein, and his 'unification' of space and time. We in fact are used to this type of visualization of time as a dimension, as we often use it to draw graphs of the behaviour of various quantities in time – the downward curve of a recession is sadly all too familiar! In Special Relativity, we draw time as an extra axis, and have rather bizarre rules for changing our velocity, called Lorentz transformations, which tell us how space and time are related.

Einstein's theory of General Relativity then does a rather sneaky thing: it curves those dimensions of space and time. When thinking of the curve of a thrown ball in the Earth's gravitational field, Einstein realized that these apparently curved paths were in fact inertial, or straight. This meant that the space around the Earth must be curved! General Relativity relates matter to curvature, and then inertial motion in this curved space translates to gravity. By doing this in a mathematically consistent way (and it took Einstein several years to assimilate the mathematics he needed), he correctly reflected the fact that gravity is just another sort of acceleration, but one in which tidal forces are real.

The largest possible canvas for Einstein's relativity is of course our universe. When we apply general relativity to the universe as a whole, we obtain a surprisingly simple model for the universe, which turns out to have one of three basic shapes, which grow in time. These shapes are a flat infinite space, a three-dimensional sphere, and what is known as a hyperbolic space. The universe is completely determined by only one varying quantity: the scale factor $a(t)$. This scale factor tells us how big the universe is, or was, at a certain time. Moreover, $a(t)$ satisfies a relatively simple equation from which a great deal of information may be inferred.

The main features of our universe turn out to be that it is dynamic (it tends to expand or contract), hotter at earlier times (the temperature drops along with expansion), and that it had a beginning, which we now call the *Big Bang*. While we cannot concretely describe the beginning (yet – another story!), we can describe the effect of the temperature evolution on the constituents of the universe.

The main huge success of the Big Bang model, and the principal reason it is the accepted model of the universe, is *nucleosynthesis*, or the process in which light nuclei are formed. We know that the universe comprises roughly a quarter helium, together with smaller abundances of other light elements such as lithium, and helium-3 (a lighter isotope of helium). Most heavy elements are synthesized in stars, but there is not enough time for this large percentage of helium-4 ('normal' helium) to have been produced in stellar cores. It must therefore have been formed in the early, hot, universe.

Fortunately, nuclear physics tells us which reactions can produce helium-4, and also how to calculate the reaction rates to compute the proportion of helium in the universe. Helium is formed in a chain of reactions, in which protons and neutrons combine to form deuterium (heavy hydrogen) and thence helium. It turns out that there is a subtle interplay between the rate at which neutrons get bound, and their decay rate (for the neutron is not a stable particle, and it decays into an electron, proton and antineutrino with a half-life of around 15 minutes). In order to get the proportion of helium we observe in our universe, we need to delay the formation of deuterium so that some neutrons can decay. This means that we need a huge bath of photons around to retard this nuclear reaction.

Alpher, Bethe and Gamow predicted this 'microwave background' – the afterglow of the Big Bang – back in the 1940s, although it was not seen until 1965 by Penzias and Wilson. The observation of this radiation background is what makes us sure that the general description of the Big Bang is correct.

We can try to apply the same ideas to calculating the abundance of matter in the early universe, which should have resulted from some earlier, higher energy reaction. However, within the Standard Model, we find we cannot. *A priori*, we expect as much matter as antimatter at the Big Bang, so we need to explain how an excess of matter was created. The problem is that any thermal process will produce equal amounts of matter and anti-matter, since they have equal mass and equal (but opposite) charge. Sakharov summed up this problem with three conditions, stating that we needed a theory of particle physics which allowed the amount of matter to change, and violated underlying symmetries, as well as a period in the early universe that was out of thermal equilibrium. So far, in spite of a great deal of effort, we do yet have a scientific theory of *baryogenesis*, the creation of matter.

Thus we have come full circle. From the ancient Greeks applying their ideas of aesthetics and deduction to the natural world around them, through the renaissance of scientific measurement, the development of the mathematical tools to describe nature, the explosion (unfortunately rather too literal) of our understanding of the very small scale and quantum nature of our world, we have arrived once again at the edge of our testable knowledge. However, we have a far better understanding of the universe, and of how it came to evolve into the rich structure we see around us today. We also have many ideas and theories which take us beyond the Standard Model, some of which include Einstein's gravity as part of their goal. However, we are stymied in our progress of picking out the correct theories because, like the ancient Greeks, our reasoning has taken us beyond our capability of testing the ideas it has generated.

In my description of the physics behind our theories of matter I have tried to explain not only the physics, but also to give an insight into the flow of ideas or the way in which the scientific community has grappled with the issues it faced. Most of science is about finding a concrete truth, yet our motivation for seeking out facts can sometimes be more an act of faith. So perhaps Einstein was right when he said: 'Science without religion is lame, religion without science is blind'.

CHAPTER THREE

RE-CREATION:
A POSSIBLE INTERPRETATION
OF QUANTUM INDETERMINISM

M. B. ALTAIE,
YARMOUK UNIVERSITY, JORDAN

Introduction

Quantum indeterminism is one of the main pillars of the argument for quantum divine action. This argument exploits the fact that events in nature can only be determined with a limited certainty, and that the occurrence of these events is probabilistic in nature. Christopher Lameter (2005) investigated this subject thoroughly with the aim of justifying belief in a God who can act in the world through a consideration of the scientific framework of quantum mechanics. Since a concept of divine action is especially relevant to theology, Lameter believes that a theory of divine action compatible with contemporary physics is a fundamental requirement for a credible consideration of how God could act in the framework of a contemporary worldview.

It is a common understanding among physicists that quantum measurement is still a problem that requires a solution in order to clarify the deep implications of quantum theory. There is no consensus among physicists: instead, we have many different views regarding how quantum measurement can be interpreted. Quantum measurement is the backbone of applied quantum mechanics, and it will be necessary to resolve this problem for the further development of quantum theory.

In this paper I will present a new interpretation of quantum measurement based on the notion of a continuous re-creation of the physical properties of systems. Using this, I will try to explain some of the basic principles of quantum mechanics on a conceptual level only, avoiding mathematical details. Then I will try to foresee some of the physical implications of such an interpretation, and will glance also upon its philosophical and theological implications.

Early development of quantum theory

The discovery of the wave properties of particles, the particle properties of waves, and the discreteness of many observables in the atomic realm, has established the need for a new description of entities in the microscopic world. At the beginning of the twentieth century many basic problems in atomic physics were addressed, leading to the establishment of quantum mechanics as a paradigm to explain the observed properties of the atomic realm. The most fundamental notions of early quantum mechanics were based on the assumption that particles behave like waves. The main difficulty in realizing a wave-like description for the particles lays in the fact that particles are localized, whereas waves are extended. This problem was overcome by Louis de Broglie's suggestion that a particle can be represented by a plain wave which has a wavelength inversely proportional to its momentum. This notion was soon utilized to obtain a description of particles in terms of a de Broglie wave-packet with the wavelength being that of the group of waves representing the particle. This description opened the way to formulating classical localized particle mechanics in terms of wave mechanics. Accordingly, a wave equation was devised by Erwin Schrödinger in 1926 to describe the time development of atomic particles under field of force (Schrödinger 1926). The need to consider the spin of the electron, and the Lorentz invariant equation of motion, required the introduction of a special relativistic formulation of the problem, and led to the well-known Dirac equation for the electron, which was discovered a few years later (Dirac 1928).

In essence, the wave-like description of atomic particles demonstrates that these display all the properties of wave phenomena, and it was soon realized that the microscopic world has some basic properties that make it different from the macroscopic world. Particles like atoms and electrons are now identified as 'quantum states', symbolized by the wave function $\psi(x,y,z,t)$. This is a mathematical expression summarizing the physical content of a physical system in terms of spacetime coordinates and other parameters of the system, like energy and momentum. The mathematical nature of $\psi(x,y,z,t)$ was recognized from the early days of formulating the Schrödinger equation, and it was realized that the wave function has no direct physical meaning in itself. But soon Max Born (Born 1926) was able to identify

$$|\psi^*(x)\psi(x)| = |\psi(x)|^2$$

as standing for the probability density[1] of finding the particle in the position x.

The wave-mechanical description of particles set by Schrödinger was best realized by saying that a particle is a wave-packet, which is composed by superposing many basic (plain) waves. This description soon faced many difficulties. The slightest dispersion in the medium will pull the wave-packet apart in the direction of propagation, and even without such dispersion it will always spread more and more in the transverse direction. Because of this blurring, a wave-packet does not seem to be a very suitable way in which to represent a particle. Shortly before Schrödinger had formulated his wave equation, during the early summer of 1925, Werner Heisenberg (Heisenberg 1925) conceived the idea of representing physical quantities by sets of complex numbers. This was soon elaborated by Born, Jordan and Heisenberg himself (Born et al. 1926) into what has become known as 'matrix mechanics', the earliest consistent theory of quantum phenomena.

Both views, the wave mechanics of Schrödinger and the matrix mechanics of Heisenberg, are said to be equivalent, despite differences in some basic concepts and formulation. A few years later Jon von Neumann (von Neumann 1955) showed that quantum mechanics can be formulated as a calculus of Hermitian operators in Hilbert space. The wavefunction was represented by complex vector in an infinite-dimensional space covered by basis vectors. According to the formalism set by von Neumann, a physical system is completely described by a state function $|\psi\rangle$, which is to be taken as a vector in an infinite-dimensional Hilbert space. A measurement of any observable a belonging to the system is the result of the action of a mathematical operator $\hat{A}$ corresponding to that observable, on the wavefunction representing the system. The result of such an operation is to produce a value (a number called the eigenvalue) that stands for the observable at the moment of measurement. On this new understanding, natural objects which were objectively identified as ontologically existing things came to be seen as new epistemological entities that are represented by abstract mathematical forms. It should be emphasized that this is a very important turning-point in the history of scientific thought. The fact that $|\psi\rangle$, which represents the physical system, is a mathematical expression that has no direct physical meaning (as noted

---

[1] The star (*) symbolizes the complex conjugate, and the total probability is the integral over all the allowed space.

earlier), and the fact that physical observables became obtainable in the theory only as a result of applying certain mathematical operators to $|\psi>$, are surely clear indications of the fundamental turn that was implied by quantum mechanics.

The eigenvalue of an operator cannot be taken as such to stand for the physical value of the observable; it has to averaged within the state of the system, and it is then called the expectation value of the operator $\hat{A}$ at the state $|\psi>$. This is the average value of all possible measurements that can be carried in the system in the state $|\psi>$.

The Heisenberg uncertainty principle

Much to the curiosity of physicists, some aspects of the wave-like description of particles led to uncertainties in determining simultaneously pairs of observables, like position and momentum, or energy and time. This was expressed by the Heisenberg uncertainty principle, which in one of its forms states that the position of a particle and its momentum can never be determined simultaneously with absolute accuracy. This principle contributed to the indeterminacy of the quantum world, and has provoked much attention and interest amongst physicists. The uncertainty principle is deeply rooted in the wave-mechanical description of particles: once we represent a particle by a wave, then it is inevitable that we should allow for some kind of a distribution of the position and momentum. The Fourier analysis of such a description shows that the wave description requires some inevitable non-locality in position, which leads to the inherent uncertainty in these variables. A similar situation arises in the measurement of time, where it leads to mutual uncertainty between time intervals and the corresponding energies. The uncertainty relations are related to the non-commutativity of the respective operators.

In matrix mechanics, operators are matrices; and in such cases $a \times b \neq b \times a$. For this reason, the operators of position and momentum do not commute, and the same is true of the operator for time and the Hamiltonian, which is the operator for energy. This in turn eliminates the possibility of finding a simultaneous eigenfunction for the position and momentum. Instead we relate the two separate eigenfunctions by a Fourier transform. It is important to note that the indeterminacy of position and momentum caused a tremendous shock to classical physics. The classical equation of motion of a particle requires knowing both the initial position and the initial momentum. Having been denied such knowledge, physicists could not solve the classical equation of motion, and this caused the downfall of classical mechanics in the microscopic world. The glory of classical mechanics, especially in its most sophisticated form devised mainly by Lagrange and Hamilton, still provokes some physicists to attempt to re-establish the reign of classical physics.

Discreteness and continuity

The quantum indeterminacy problem is deeply rooted in the long-lasting question of discreteness and continuity. This is an issue which has been under persistent debate since the early days of the Greeks, continuing throughout the Islamic period which witnessed fierce debates between the philosophers and the Mutakallimūn (for a detailed account of Kalām atomism see Wolfson 1976). In this context, it is noteworthy that the Mutakallimūn adopted profoundly the notion of re-creation on a conceptual level, and it led them to recognize that natural events should be indeterministic (see Altaie 2007).

The indeterminacy of quantum states, as described by the Heisenberg uncertainty principle, brought to the attention of physicists the fact that quantum mechanics is a mechanics of an undetermined nature. As noted above, this soon posed what came to be known as the 'measurement problem' in quantum mechanics. Today, more than three quarters of a century after the advent of the theory, this is still an issue of unprecedented debate. In fact, it is by far the most controversial problem of current research in the foundations of physics, and it divides the community of physicists and philosophers of science into numerous opposing schools of thought. The main issues in this division seem to be centered a round two things: quantum jumps, and measurement indeterminacy.

Quantum jumping is an indication of the discrete nature of the atomic world. If this is a fundamental characteristic of the microscopic world, then the perceived continuity of the macroscopic world would seem to be illusory. It was reported that Schrödinger once said: 'If all this damned quantum jumping were really to stay, I shall be sorry I ever got involved with quantum theory' (Jammer 1974, p. 57). The main difficulty arises when we find differential calculus (which is the backbone of the mathematical formulation of classical physics, based on continuity and infinite divisibility) to be in need of serious revision. Consequently, the canonical formulations of physical laws will not be valid, and the basic concepts of field theory will be challenged. The Schrödinger equation is a deterministic equation that adopts the principle of continuity and the concept of infinite divisibility. However, it is also a wave equation, that has helped to provide an approximate picture of the quantum world. The discrete features of the quantum world are now being presented as products of its wave mechanical nature, which allows for the superposition of waves producing an interference pattern. Consequently, one can avoid thinking in terms of abrupt quantum jumps in favour of thinking in terms of probability distributions, such that some kind of continuity between discrete states is maintained. In this way, instead of having macroscopic continuity becoming an apparent feature that hides the underlying discreteness, we now have discreteness appearing as an emergent product of some phenomena of the continuum. In addition to this, it is important to note that precise analysis of the quantum phenomena of the two slit interference experiment shows some fundamental characteristic departures from the standard wave-interference phenomenon (Namiki and Pascazio 1993). In these experiments a particle continues to be non-divisible; but at very low intensities the behaviour of a particle beam may be shown to be different from the behaviour of photons in an electromagnetic beam. Such divergence of behaviour awaits an explanation which can precisely identify those features in both phenomena that make them different.

The applicability of quantum mechanics

In this context comes the question of whether quantum mechanics is a theory that can be applied to a single particle, or whether it is a theory of ensembles. Physicists have different opinions on this issue. Some of them, like Bohr and Heisenberg, believe that quantum mechanics is suitable for describing both single particles and many-particle systems. This is generally the view held by the Copenhagen school. Other physicists, like Einstein and Born, believe that quantum mechanics is only applicable to ensembles rather than to individual particles, and accordingly maintain that it can only be interpreted statistically. Still others, like Everett and Wheeler, believe that quantum mechanics is essentially an interaction theory, which can be realized only through the interaction between the observer and the system. In one way or another, this interaction allows for a subjective interference in determining quantum states. In fact, the basic formulation of the equation of motion in quantum mechanics, Schrödinger's equation, suggests that it can be applied to single particles. On the other hand, having the values of observables coming out as an average only may suggest that we are talking about an ensemble of particles, in which each particle enjoys a different value for a given observable. The general behavior of the system of these particles is then represented by the behaviour of the average. However, this restriction becomes unnecessary if we interpret the existence of an average as occurring as a result of many measurements being performed on the same particle. In this case, the implicit fact will be that the value of the observable assigned to the system (the single particle in this case) is not fixed, but is ever changing. But then the question arises as to whether this change in the value of the observable is due to the changing state of the system, or whether it is due to the process of measurement itself. If we assume that it is due to the changing state of the system, then the process of measurement can be taken to be completely passive. On the other hand, if we consider it to be a result of the measurement itself, then we are assuming primarily that the measurement has a disturbing effect on the system. This amounts to assuming the existence of an interaction between the system and the measuring device. Since the microscopic systems under investigation are so small and delicate, no-one can deny that such interactions are possible, and may cause subsequent disturbances. Therefore, such interactions will lead to the 'decoherence' of the quantum system. The disturbances caused by the measuring devices are generally non-systematic, and are so complicated that they would be unpredictable. On the other hand, one might expect that in some cases the disturbances caused by the macroscopic measuring device would be so large as to overwhelm the basic value of the observable under measurement.

A third point to be made here is that such disturbances, if known, can be accounted for in the equation of motion through the potential term in that equation. Accordingly, the case will always be that of an interacting system for which the equation of motion may be solved exactly or through numerical techniques. Virtually any environmental factor can be included in the potential of the system, which controls the system's behaviour through the equation of motion. Taking these points into consideration, it would be odd to assume that quantum indeterminacy is simply a result of the imprecision of measurement.

Interpretations of quantum measurements

In a given individual experiment, the result of the measurement is one of several alternatives. A repetition of the experiment under identical initial conditions may lead to another of these possible alternatives. This is incompatible with the unitary evolution of Schrödinger. Several solutions have been proposed for this apparent inconsistency. The main ones are:

1. The von Neumann interpretation: wavefunction collapse.
To explain the process of measurement, von Neumann suggested that the state function changes, in two different ways (c.f. von Neumann 1955):
Process 1: a discontinuous change brought about by the observation by

which the quantity with eigenstate $|\psi>$ is projected on the state $|\varphi> = \hat{A}|\psi>$ instantly, with probability $|<\psi \mid \varphi>|^2$. This amounts to determining the overlap between the state $|\psi>$ and the state $|\varphi> = \hat{A}|\psi>$.

Process 2: a change in the course of time development, according to the deterministic Schrödinger equation.

The description in process 1 is called 'the wavefunction collapse'. This means that the state $|\psi>$, after measuring the observable a, will be converted into the state $|\varphi> = \hat{A}|\psi>$.

A fundamental problem was recognized long ago in this formulation of von Neumann. This problem is the apparent inconsistency between the indeterministic nature of process 1 and the deterministic nature of process 2. This apparent inconsistency has been presented in different forms, and it is in fact deeply rooted in the formulation of quantum mechanics from its very beginning. Josef Jauch (Jauch 1973) presented the problem as follows: the problem of measurement in quantum mechanics concerns the question of whether the laws of quantum mechanics are consistent with the acquisition of data concerning the properties of quantum systems. This consistency problem arises because the system to be measured, and the apparatus which is used for the measurement, are themselves systems which are presumed to obey the laws of quantum mechanics. The evolution of the state of such a system is therefore governed by the Schrödinger equation. However, the measuring process exhibits features which are apparently inconsistent with Schrödinger-type evolutions. The typical process ends with the establishment of a permanent and irreversible record, and this contradicts the time-reversible Schrödinger equation. So, despite the fact that the von Neumann interpretation of quantum measurement was adopted by the Copenhagen school, it nevertheless suffers from some fundamental problems.

2. The statistical interpretation.

For this we have two views:

Viewpoint 1, by which quantum mechanics is understood to apply to ensembles and not to single particles. Albert Einstein was an advocate of this interpretation. Einstein says: 'The function ψ does not in any way describe a condition which could be that of a single system: it relates rather to many systems, to "an ensemble of systems" in the sense of statistical mechanics' (Einstein 1936). Einstein hoped that a future, more complete, theory might describe quantum mechanics as an approximation of a more general one.

Viewpoint 2, which was proposed by Born and supported by Bohr, according to which the wavefunction ψ was understood to be a symbolic representation of the system, and $|\psi(x)|^2 = |\psi^*(x)\psi(x)|$ is taken to describe the probability density for the system in the position x. But probability can only be understood to have a meaning through a population. In this case, the population is that of many repeated measurements. This may be asserted by the fact that Born was of the opinion that his suggestion has the same content as that of Einstein, and that 'the difference [in their views] is not essential, but merely a matter of language' (Born 1971, p.10).

One can say that the Einstein interpretation is covered by the fact that in any measurement on a quantum system we measure macroscopic quantities, a fact which was originally emphasized by Bohr. If, however, we come to measure by any means a microscopic quantity, then the Einstein interpretation will not be valid. On the other hand, by requiring that many measurements are to be done on the same system, Born's interpretation implicitly assumes that the system is to remain within the same state over the duration of all those measurements. Obviously this cannot be generally guaranteed.

3. The hidden variables interpretation.

This interpretation was championed by David Bohm (Bohm 1952a, 1952b), who assumed that quantum mechanics is incomplete, and that there are some hidden variables that should complement the physical description in order to obtain the full picture of the physical world, which is assumed to be deterministic. There are several kinds of hidden variable theories: some are local and some are non-local. Belinfante has given a very detailed account of these theories both in their scientific content and in their historical development (Belinfante 1973). By Bell's theorem (Bell 1964), the local hidden variable theories were shown to be inconsistent with quantum mechanics. It may be added that none of the existing non-local theories makes any prediction that is new to the standard formulation of quantum mechanics.

4. The multi-world interpretation.

This was originally proposed by Hugh Everett III (Everett 1957). Everett reformulated the process of measurement, abandoning the concept of wavefunction collapse set out in process 1 of the von Neumann formalism, while keeping the assumption of the deterministic evolution of the system under the Schrödinger equation. Everett criticized the need for 'external observers' to obtain measurements by the von Neumann scheme, and instead considered the system as being composed of two main subsystems: the object and the measuring device (or observer). This formulation established the concept of 'relative state'. The treatment led Everett to conclude:

> throughout all of a sequence of observation processes there is only one physical system representing the observer, yet there is no single unique state of the observer (which follows from the representations of interacting systems). Nevertheless, there is a representation in terms of a superposition, each element of which contains a definite observer state and a corresponding system state. Thus, with each succeeding observation (or interaction), the observer state "branches" into a number of different states. Each branch represents a different outcome of the measurement and the corresponding eigenstate for the object-system state. All branches exist simultaneously in the superposition after any given sequence of observations (Everett 1957).

Everett went further, to suggest that:

> the trajectory of the memory configuration of an observer performing a sequence of measurements is thus not a linear sequence of memory configurations, but a branching tree, with all possible outcomes existing simultaneously in a final superposition with various coefficients in the mathematical model. In any familiar memory device the branching does not continue indefinitely, but must stop at a point limited by the capacity of the memory (Everett 1957).

John Wheeler supported the Everett theory, emphasizing its self-consistency (Wheeler 1957). An elaboration of the Everett interpretation was also the subject of a study by Graham, working under the supervision of Bryce DeWitt (Graham 1970). It was assumed that the eigenvalues associated with the observer subsystem form a continuous spectrum, whereas the eigenvalues associated with the object form discrete set; and in order to reconcile the assumption that the superposition never collapses with ordinary experience, which ascribes to the object system after the measurement only one definite value of the observable, it was proposed that the world splits into many worlds existing simultaneously. In each separate world a measurement yield only one result, though this result differs in general from one world to another.

The re-creation postulate

In order to interpret quantum measurement, I propose the following two postulates:

Postulate P(1): All physical properties of a system are subject to continuing re-creation.

Postulate P(2): The frequency of re-creation is proportional to the total energy of the system.

It will be shown below that the re-created observables assume new values every time they are re-created. This will cause the observables to have a distribution of values over a certain range (width) that is always controlled by the re-creation frequency. The higher the total energy of the system, the narrower is the range of values over which the dispersion is expected (and vice versa). For this reason, macroscopic systems are expected to behave classically, whereas microscopic systems exhibit mostly quantum behavior. Clearly, the narrower the dispersion of values, the more determinable is the value of the observable, and vice versa.

Re-Creation and the Uncertainty Principle

Once created, an observable assumes a given basic value defined by the state of the system at that moment. According to the re-creation postulate P(1), physical parameters are in a natural process of continuing re-creation, irrespective of the measurement operation. However, the values of those parameters can only be known at the time of measurement. Re-creation is a process of change. Once a given parameter is re-created other parameters of the system will be affected, thus changing their values in accordance with the related physical laws. Any change is best described in the most general form by the generator corresponding to that parameter. For example, if the position x is re-created then the system will change infinitesimally by $\partial/\partial x$, but this is simply proportional to the momentum operator. This will duly cause the value of the position x to change every time it is re-created, thus presenting a distribution of values for x instead of one single value. Conversely, if the momentum p is re-created then the whole system will change by $\partial/\partial p$, but this will cause an infinitesimal shift in the value of the momentum p and consequently a shift in the value of the position parameter x. Therefore every time a position x is re-created a change in the momentum of the system will occur; and, conversely, every time the momentum p is re-created a change in the value of the position will occur. This means that re-creating the position will result in creating momentum and vice versa. If the system itself is to stay invariant under the process of re-creation then we must have

$$\left(\frac{\partial}{\partial x}x - x\frac{\partial}{\partial x}\right)|\psi> = |\psi>$$

Using the explicit forms for the position and momentum operators, this would imply that

$$\hat{p}\hat{x} - \hat{x}\hat{p} = [\hat{p}, \hat{x}] = -i\eta$$

In other words, the effect of change is logically being seen as a commutation of the parameter and its generator (which were also called complementary observables). This is the well-known commutation relation that led to the Heisenberg uncertainty relations. In this scheme, however, measurements could be passive actions that do not necessarily affect the system itself.

This proposal of re-creation preserves the statistical nature of the possible values of the observables, and resolves the question of whether quantum mechanics is applicable to a single particle or to an ensemble of particles. Here we see that the single particle state is being continually re-created, thus forming an ensemble of values *on its own*, if a memory is to be available to keep records of all the values assumed under re-creation. Nevertheless, a measurement of an observable, taken over a period of time exceeding the re-creation period, will always yield an average of the values assumed by the system during that period of measurement. So, in practice we measure average values every time we perform a measurement. This explains how probabilistic behaviour arises in the case of a single particle quantum system. According to the above scheme, we always measure average values with very low dispersion for macroscopic objects: the re-creation frequency is very high, and consequently the measurement time cannot cope with the re-creation period. This gives the macroscopic world its classical, apparently deterministic, characteristics. This is why the measured values of the observables of a macroscopic system are always very close, even identical, to the theoretically-expected values of the observables. On the other hand, in microscopic systems the re-creation frequency is relatively low, and therefore we would expect the dispersion of values to be high enough to reveal the indeterministic character of the world.

This proposal also provides us with a better understanding of the origin of the uncertainty relations. Here we see that the appearance of uncertainty in the values of complementary observables is a direct result of re-creation, and of the entanglement of such variables. This means that indeterminism is a direct consequence of continuous re-creation.

Physical Implications of Re-creation

There are several implications following from the proposed re-creation scheme described above. Some of these implications might be used to test the theory. However, because of the largely technical nature of these implications, I will only provide an overview of those implications that might be of interest to people working on issues in the science and religion debate. The full technical treatment of these implications will need to be presented elsewhere.

1. Macroscopic quantum states.
The re-creation frequency can be affected by an external field of force. Since it is known from the theory of general relativity that any time duration for an event occurring near a gravitational field of force is dilated by a factor proportional to the strength of the field, then one should expect that re-creation periods will be dilated when they are in the vicinity of a strong source of gravity (cf. Weinberg 1972). Consequently, re-creation frequencies should be red-shifted when they are in the vicinity of a strong gravitational source. This means that macroscopic classical processes would start to exhibit quantum features when they are in a strong gravitational field. This will cause the appearance of macroscopic quantum states in such regions, for example near the event horizon of black holes.

2. Quantum coherence.
Coherence is a phenomenon in nature whereby efficient transformation of energy takes place. In the macroscopic world, coherence is shown when two oscillating systems resonate with same frequency and are in phase: for example, a system comprising two adjacent pendulums. However, we do not usually talk about such resonating systems in the macroscopic world as being coherent states. This phenomenon is best realized in quantum systems. Such systems always display high efficiency in energy transformation (for example, lasers). The availability of macroscopic quantum states may make it plausible to expect the occurrence of macroscopic coherent states too, thus opening the way to understanding some very obscure phenomena, like the gamma-ray bursts which are known to occur at the far rim of the universe. In addition to this, the re-

creation postulate allows for a new definition of coherence, by which two systems can be considered to be in a coherent state if their re-creation frequencies are identical, and their re-creation happens to be in phase.

3. Quantum Zeno Effect.

The Quantum Zeno Effect (QZE) is a very interesting proposal which was suggested by Misra and Sundarshan (Misra and Sundarshan 1977). Their proposal is based on the notion of wavefunction collapse, and was considered to be a prediction of the collapse interpretation. The idea is that if continuous measurements are carried out on a given state, then the system is expected to stay in that state because of the continuous collapse of the wavefunction onto the same state. (As they say, a watched pot never boils!) The verification of this prediction has been claimed (Itano et al. 1990), but such claims have been refuted (Petrosky et al. 1991). Recently, some more rigorous calculations have been made attempting to present the QZE quantitatively in more accurate form by taking into consideration the effect of the measurement duration (Schulman et al. 1994, Schulman 1997). The re-creation interpretation presented in this paper sets an upper limit for the measurement time for the QZE to be verified. The measurement time of an observable (say, transition energy) should be less than the re-creation period for the QZE to occur. Measurements performed within a time duration which is more than the re-creation time will result in averaging the values of the observable over several re-created states, and consequently cannot hold the system at a specific state. Consequently the QZE will not be verifiable if the measurement time is more that the re-creation time.

Conclusion

The scheme proposed in this paper for the interpretation of quantum indeterminism offers a scope that allows for an objective ontology of the physical world in addition to the possibility of its being undetermined. Such a scheme is more realistic and more consistent than the observer-dependent interpretation which is implied by the von Neumann and the Everett-Wheeler interpretations. The re-creation scheme is free of the known paradoxes of quantum measurements, like Schrödinger's cat and the EPR paradox, since it does not consider a subjective role for measurements or a wavefunction collapse, whilst it does assume the natural presence of entanglement in states belonging to the same system. Moreover, this scheme resolves the statistical nature of quantum mechanics by allowing the statistical distribution of the possible values that an observable might take to fall within the natural process of continued re-creation of that observable.

It is important to note that the above scheme will not affect the standard calculations of quantum mechanics, except that it might motivate new investigations into regions which until now have not been explored by mainstream research. These could include the existence of macroscopic quantum states, and the possibility of understanding gamma ray bursts in terms of some macroscopic quantum processes taking place under very specific conditions deep in the universe. However, the scheme proposed here is by no means complete, and is open to further development. The technical part of this proposal, including all the mathematical details, will be published elsewhere.

Acknowledgment

This work was supported by a grant from the John Templeton Foundation.

Bibliography

Altaie, M. B. 2007. 'Islamic Kalām: a Possible Role in Contemporary Science and Religion dialogue', Annals of the Sergiu Al-George Institute of Oriental Studies XI, 49-57.

Belinfante, F. J. 1973. A Survey of Hidden Variables Theories. Oxford: Pergamon Press.

Bell, J. S. 1964. 'On the Einstein Podolsky Rosen paradox', Physics 1, 195-200.

Bohm, D. 1952a. 'A Suggested Interpretation of the Quantum Theory in Terms of "Hidden" Variables, I'. Phys. Rev. 85, 166-179.

—. 1952b. 'A Suggested Interpretation of the Quantum Theory in Terms of "Hidden" Variables, II'. Phys. Rev. 85, 180-193.

Born, M. 1926. 'Zur Quantenmechanik der Stoßvorgänge', Zeitschrift für Physik 37, 863-867.

Born, M., Heisenberg, W. and Jordan, P. 1926. 'Zur Quantenmechnik II', Zeitschrift für Physik 35, 556-615.

Born, M. 1971. The Born-Einstein Letters. New York: Walter and Co.; London: Macmillan.

Dirac, P. A. M. 1928. 'The Quantum Theory of the Electron', Proc. Roy. Soc. (London) A 117, 610-624.

Einstein, A. 1936. 'Physics and Reality', Journal of the Franklin Institute 221, 313-347.

Everett, H. 1957. 'Relative State Formulation of Quantum Mechanics', *Rev. Mod. Phys.* 29, 454-462.

Graham, N. 1970. 'The Everett interpretation of quantum mechanics', Ph.D. thesis, University of North Carolina at Chapel Hill.

Heisenberg, W. 1925. 'Über quantentheoretische Umdeutung kinematischer und mechanischer Beziehungen', *Zeitschrift für Physik* 33, 879-893.

Itano, W. H., Heinzen, D. J., Bollinger J. J. and Wineland, D. J. 1990. 'Quantum Zeno Effect', *Phys. Rev.* A 41, 2295 – 2300.

Jammer, M. 1974. *The Philosophy of Quantum Mechanics: The Interpretations of Quantum Mechanics in Historical Perspective.* New York: John Wiley & Sons

Jauch, J. M. 1973. 'The Problem of Measurement in Quantum Mechanics', in J. Mehra (ed.), *The Physicist's Conception of Nature.* Boston: D. Reidel Publishing Company.

Lameter, C. 2005. *Divine Action in the Framework of Scientific Knowledge.* Newark, California: Christianity in the 21st century.

Misra, B. and Sudarshan, E. C. G. 1977. 'The Zeno's paradox in quantum theory', *J. Math. Phys.* 18, 756–763.

Namiki, M. and S. Pascazio, S. 1993. 'Quantum Theory of Measurement Based on the Many-Hilbert-Space Approach', *Physics Report* 232, 301-411.

Petrosky, T., Tasaki, S. and Prigogine, I. 1991. 'Quantum Zeno Effect', *Physica* A 170, 306-325.

Schrödinger, E. 1926. 'Quantisierung als Eigenwertproblem' *Ann. Phys.* 79, 361-376.

Schulman, L. S., Ranfagni, A. and Mugnai, D. 1994. 'Characteristic scales for dominated time evolution', *Physica Scripta* 49, 536-542.

Schulman, L. S. 1997. 'Watching it Boil: Continuous Observation for the Quantum Zeno Effect', *Found. Phys.* 27, 1623-1636.

von Neumann, J. 1955. *Mathematical Foundations of Quantum Mechanics*, trans. Robert T. Beyer. Princeton, New Jersey: Princeton University Press

Weinberg, S. 1972. *Gravitation and Cosmology.* New York: John Wiley & Sons.

Wheeler, J. A. 1957. 'Assessment of Everett's 'relative state' formulation of quantum theory', *Rev. Mod. Phys.* 29, 463-465.

Wolfson, H. 1976. *The Philosophy of the Kalām.* Cambridge Massachusetts: Harvard University Press.

PART II:

HISTORICAL PERSPECTIVES ON MATTER

CHAPTER FOUR

THEOLOGY AND MATTER THEORY IN THE EARLY MODERN PERIOD

PETER HARRISON,
UNIVERSITY OF OXFORD

… it seems probable to me, that God in the Beginning form'd Matter in solid, massy, hard, impenetrable Particles, of such Sizes and Figures, and with such other Properties, and in such Proportion to Space, as most conduced to the End for which he form'd them (Newton 1952, p. 200).

Matter, appearance and reality

The basic assumption underlying any theory of matter is that beneath the manifold appearances of the world as it presents itself to our senses, there lie more fundamental, unseen realities. While the distinction between appearance and reality is perhaps most familiar from the discussions in Plato's *Parmenides*, in fact it informed the very earliest natural philosophical speculations of the Greeks. Among the presocratic philosophers, various material principles were postulated as the basic stuff of material reality: water (Thales), air (Anaximenes), fire (Heraclitus), the four elements – earth, air, fire, water (Empedocles). More abstract, non-material, notions were also proposed as the underlying reality of all things – 'the boundless' [*apeiron*] (Anaximander) and 'mind' [*logos*] (Heraclitus). For each of these thinkers, the world as it appears to us in all its variable and multiple guises is, in reality, just one kind of unchanging thing.

Among the speculations of the presocratic philosophers, one was to have enduring importance. In the fifth century BC Democritus proposed that matter is constituted at its most fundamental level by indivisible particles called atoms which move in the void. Atoms thus represented the unchanging substrate, their various sizes and shapes, motions and combinations, providing the changing phenomena that make up the world of appearances. Subsequently Epicurus and Lucretius were to adopt this atomistic philosophy. From the perspective of later Christian thinkers, however, atomism had little going for it: it was based on chance and necessity, it ruled out providence, and its ethical implications were unwelcome. Partly for these reasons, virtually from the time of its conception, atomism was regarded as incipiently atheistic. More congenial from the standpoint of patristic and medieval theologians were the views proposed respectively by Plato and Aristotle. (And, as we shall see, Stoic ideas also proved to be influential).

Plato's view, as is well known, was that the shadowy and evanescent objects of the sensory world were but imperfect and transient copies of the ideal objects in the world of the forms. Although the material world was real, it was inherently less perfect than the immaterial world of ideas. Reflecting this dichotomy, human beings were made up of a soul, the rational aspect of which originated from the world of forms, but which in this earthly existence was imprisoned in a material body. Given this view of the nature of things, the goal of the philosophical life was to liberate the soul from its material prison, so that upon death the soul could return to the unseen world from which it had originated. This position also had significant epistemological implications. Hence, for Plato, there could be no genuine knowledge of the ephemeral and constantly changing objects of sense. Accordingly, true *scientia* required knowledge not of specific material things, but of the eternal forms in which material things participated. This knowledge was to be achieved by shutting out the distracting images of material things derived from the senses, and intuiting or recollecting the pure forms that lay within the mind. These ideas, combined with those of the Pythagoreans, also promoted the idea that universal and unchanging mathematic realities underlie the structures of the cosmos.

Plato's most famous pupil, Aristotle, agreed with his teacher that the forms of things were important and that the essential natures of things are determined by their form, but he denied that these forms had an existence that was independent of particular material instantiations. For Aristotle, forms were not ideal, existing in their own separate world, but were 'substantial' or 'accidental'. Aristotle articulated the intuitive idea that objects are to be understood in terms of form and matter. If we consider a bronze statue, for example, we can easily understand that its matter is the bronze, and its form the particular shape which it takes on. Aristotle also took over from Empedocles the idea that there were four basic forms – earth, air, fire and water – which could be organized into more complicated things. In the case of living beings, the form is also the source of life, namely, the soul.

For Aristotle, the concepts of form and matter together answer the question, for any particular thing, what is it? And because these concepts form part of the explanation of what something is, Aristotle will speak of 'formal' and 'material' *causes*. But there are further questions that we can ask about things, such as: how did it come to be? Or, what is its purpose? For Aristotle these questions are answered by reference to two further causes: the efficient cause and the final cause. The efficient cause – which of Aristotle's four causes is closest to our modern understanding of cause – is that which brings about motion or change. The final cause is the internal tendency of the thing to move towards its natural place, to fulfill its natural purpose, or to achieve its natural perfection. Final causation accounts for the apparently goal-directed behaviours of natural things. It is important not to conflate Aristotelian teleology with notions of divine purpose. Aristotle's god is not a creating deity who brings things into being for his own purposes. Rather, things have their own natural purpose and tendency. Aristotelian teleology, then, entails purpose, but not a divinely imposed purpose.

In summary, the ancient Greeks were more or less agreed that there was some underlying reality to things, but they disagreed on what that ultimate reality actually was.

Matter and creation in the Middle Ages

The earliest forms of Christian theology arose, at least in part, out of attempts to come to terms with the legacy of classical philosophy. St Paul's critical comments about 'the wisdom of the world', and Tertullian's well-known remarks about the opposition between Athens and Jerusalem, might imply a rejection of Greek philosophy, and certainly philosophy was perceived to have its dangers. Various forms of Gnosticism, which took over and exaggerated the Platonic notion of the material world's intrinsic inferiority, were regarded as a particular danger on account of their disparagement of a material world. Such a view had implications for the Christian doctrines of creation, incarnation and resurrection, each of which requires a rather more positive appraisal of material things. That said, most of the Church Fathers were willing to concede some value to pagan wisdom. A standard formula, worked out by the Alexandrian Church fathers, suggested that Greek learning was a 'preparation for the gospel'. This idea, taken over into medieval understandings of the relationship, held that Christianity represented the unfulfilled goal of pagan philosophy. Christianity, as Augustine was to express it, was 'the one true philosophy'.

Christian theology was in time to offer a quite distinctive idea of creation. Aristotle had contended that the world was eternal, while Plato had suggested that the world had been fashioned from pre-existing matter by the Demiurge. This latter view was held by various Gnostic thinkers, who further developed the idea that the pre-existing matter from which the world had been fabricated was recalcitrant and inherently defective. The Christian alternative to this, which began to crystallize during the patristic period, was that God had created the world *ex nihilo* – out of nothing. While this view does not receive unambiguous support from the Hebrew bible, it seemed important to a number of the Church Fathers, particularly in a context in which the Gnostic alternative was proving attractive, to stress that God's power in creation was not limited by the materials that he had to work with.[1] Augustine's doctrine of creation, although distinctive in some respects, reflects these priorities.

Those familiar with the details of Augustine's biography will know that he was particularly concerned to refute the claims of Manichaeism, a syncretistic religion which had once exercised a great attraction for him. The Manicheans taught that there existed two eternal and opposed forces, good and evil, and two realms, the spiritual realm of light and the material realm of darkness. They also contended that the biblical creation story, with its anthropomorphic notion of a six-day creation, and it clear implication that the material world was good, was a nonsensical and primitive myth. In contesting this view, Augustine was concerned both to do justice to the relevant biblical texts, and to articulate a philosophically coherent view of the origins of the cosmos. He develops his view in a number of works, the most important of which was *De Genesi ad litteram*. The first difficulty that Augustine faced was the narrative of the six-day creation in Genesis 1, a passage which had attracted the scorn of the Manicheans. Earlier commentators like Origen had simply dismissed this passage as allegory. Augustine was also to reject the idea of a six-day creation, arguing instead, partly on the basis of a verse in Sirach, that God had created everything at once.[2] This was to become the standard view of the creation throughout the middle ages.

There remained the question, however, of the meaning of the six-day creation. In order to resolve this issue, Augustine drew upon the Stoic idea of *rationes seminales* – seed-like principles that came to fruition over the development of time. Augustine's idea was essentially that God had

---

[1] The first verse of Genesis is perhaps more properly translated something like: 'In the beginning, when God was creating the heavens and the earth, the earth was without form ...' Support for the *ex nihilo* view comes from II Macabees 7:28.
[2] Sirach 18:1: in the Latin translation that Augustine used, 'He made all things together'.

created the 'seeds' of all things at once, and thus while all things were present at the beginning of time, some were so only potentially. To illustrate his point, Augustine uses the analogy of the way in which a tree is potentially present in its seed. The initial creation, he writes, thus included 'those things that water and earth produced potentially in their causes, before they could evolve through the intervals of time' (Augustine 2002, p. 300: see also Gilson 1961, pp. 197-209). Such a view was reinforced by the Genesis text which speaks of 'the earth' bringing forth living things. For Augustine, the earth was the source of hidden potentials of future living things.

The eleventh century witnessed the reintroduction of Aristotle's works into Latin West, sparking a remarkable renaissance of philosophical and scientific thinking. The work of Thomas Aquinas (c. 1225-74) exemplifies the way in which Aristotelian ideas were incorporated into medieval theology. The Greek philosopher had stressed that the immanent powers of natural things, and explanations couched in terms of those powers. Aquinas provided a theological gloss, speaking of 'the order that God has implanted in nature' (Aquinas 1934, vol. 4 p. 58: Aquinas 1932, p. 143). Divine activity, on this account, is typically manifested in nature through the operation of secondary causes, which is to say, causes proper to nature itself. Natural things thus possess their own causal potency, which springs from the kinds of things that they essentially are. Again, referring back to the problem of appearance and reality, scientific knowledge of things is a knowledge of their unchanging and distinct essences (Aristotle 1933, vol. 1 pp. 298-304, vol. 2 pp. 86-94: for mediaeval and renaissance versions of this view, see Aquinas 1961, Palmieri 1997, esp. p. 153). The best model for nature on this view (as for Augustine's, too) is that of an organism.

It has recently been suggested that the medieval alchemical tradition represented a significant alternative to Aristotelian conceptions of matter. Whereas Aristotelianism accounts for the properties of chemical compounds by making reference to their substantial forms, thinkers such as the thirteenth-century alchemist Geber claimed that matter was made up of small particles that retained their basic integrity through various chemical transitions. In essence this was an atomist theory, which some have thought was indebted to Arabic atomism. Such a view of matter was a necessary presupposition for the possibility of the transmutation of metals. Scholastics argued against this that upon combination, chemical compounds took on a completely new form – the substantial form – which comprised its specific essence. According to this view, the forms of the original constituents of any compound were lost forever. The alchemists based their views on alchemical experiments. Some historians argue, on this basis, that medieval alchemy may represent an important source for the predominant matter theory of the early modern period – the corpuscular philosophy. (On the revival of Epicureanism see Wilson 2008, pp. 1-38: for the tradition of medieval alchemy and its influence on early modern ideas and practices, see Newman 2006. See also Lüthy et al. (eds.) 2001).

The corpuscular hypothesis

The Renaissance witnessed the recovery of a range of ancient Greek texts, and with them a new familiarity with long neglected natural philosophical traditions. Recovered works included the writings of the ancient Greek atomists Democritus and Epicurus, and paved the way for a recovery of their ideas. Philosophical arguments drawn from the ancients were used to support what become known as 'the corpuscular hypothesis', a view of matter which was explicitly opposed to the prevailing Aristotelian view. As suggested above, medieval alchemical practices are also a likely source of atomistic ideas, and it is possible that Arabic atomism, as exemplified in Asharite philosophy and the thought of al-Ghazali, had an influence in the medieval West, and indirectly on early modern philosophy (Wolfson 1976: Glasner 2001, pp. 9-26).

Atomist ideas circulated in sixteenth-century England, and were popular with a group known as the Northumberland circle. Early in the seventeenth century Francis Bacon advocated atomism, while in Italy Giordano Bruno and Galileo were both atomists. However, it was the French philosopher-priest Pierre Gassendi (1592-1655) who played a vital role in reviving atomism and furnishing it with philosophical justifications. Subsequently, it was his compatriot René Descartes who first formulated a complete philosophical system based on atomism. A key feature of Descartes' view was that the minute particles that provided the most fundamental units of matter were completely inert. This was in stark contrast to the older Aristotelian idea, according to which matter had inherent powers, tendencies and goals. All of the activities in nature, on the Cartesian mechanical model, were to be accounted for in terms of the size and shape of the fundamental particles, and their motions. In principle, 'matter and motion' (Galileo's expression) were all that was required to account for the operations of nature. While there were disagreements about details of this system, this atomic theory of matter – or 'the corpuscular hypothesis', to give it its proper label – became the common matter theory of the new science.[3]

The corpuscular hypothesis went hand in hand with a new approach to nature – the mechanical philosophy – which held that the operations of nature were analogous to those of complex machine. (Not all corpuscularians were mechanical philosophers, however, and even in those philosophers thought to exemplify the unity of these views there is a distinction to be observed. Generally, mechanical philosophers insisted that corpuscles or atoms are intrinsically inert: see, e.g., Clericuzio 1990, pp. 561-589.) This conception of nature contrasted with the more organic model that had characterized medieval Aristotelianism. As Johannes Kepler wrote, the 'celestial machine' should be thought of 'not on the model of a divine, animate being, but on the model of a clock ... In [that machine] almost all the variety of motions [stems] from one most simple, physical magnetic force ... And I mean to call this form of reasoning 'physics' [done] with numbers and geometry' (Kepler to Herward von Hohenberg, 10 February 1605: see Mahony 1998, p. 706). Kepler notes that this mechanistic model of the cosmos better lends itself to mathematical analysis than the organismic model.

In certain respects, the mechanical philosophy offered an approach to the natural world that was quite compatible with religious conceptions of nature. Whereas on the organic model the organizing principles of the world are to some extent self-explanatory and attributable to 'nature', a machine needs someone to design it and to create it. (It was partly for this reason that in his criticisms of the argument from design David Hume had proposed an organic model of the world.) In this respect, the mechanical philosophy seemed ideally suited to a theistic understanding of nature, and to the idea of a creator. Moreover, the 'new' corpuscular philosophy was typically understood as a revival of an ancient biblical or 'Mosaic' philosophy. This ancient philosophy, it was assumed, had provided the source of the later ideas of the Greek atomists. The seventeenth-century emergence of mechanical philosophy was accordingly regarded as the reformation of a true natural philosophy, akin to the reformation of religion that had preceded it (Harrison 1998, pp. 138-46).

In other respects, however, the combination of corpuscular matter theory and mechanical philosophy was theologically problematic. An immediate difficulty lay in the atheistic reputation of atomism. Not only was atomism associated with atheism in antiquity, but Thomas Hobbes,

---

[3]   On a strict definition, atomism calls for indivisible particles and a void. Corpuscularians might deny the void, as Descartes did, and/or allow that particles were invisible, but not necessarily indivisible. Again, Descartes maintained that in principle God could divide anything that was extended (Descartes 1984, vol. 1).

who was one of the most consistent seventeenth-century advocates of atomism and mechanism, was widely regarded by his contemporaries (probably unjustly) as a materialistic atheist. So the corpuscular hypothesis suffered the problem of guilt by association. Advocates of atomism also encountered difficulties explaining key Christian doctrines. Transubstantiation – an explanation, couched in Aristotelian terms, of how the elements of the mass are transformed into the body and blood of Christ – became almost impossible to account for in terms of the corpuscular philosophy. In spite of his valiant efforts to do precisely this, Descartes' works were placed on the Index in 1663 on account of the apparent inconsistency between Cartesian philosophy and transubstantiation (Bourg 2001: Nadler 1988: Armogathe 1977). The question of the human soul was also a central issue. Previously understood in Aristotelian terms, the soul had been regarded as the substantial form of the person. With the demise of Aristotelian forms and formal causes, this doctrine needed to be reformulated. Descartes responded to this latter challenge by positing a soul or mind that consisted of a non-extended substance. The Cambridge Platonists revived ancient Neoplatonic ideas involving terrestrial, aerial, and ethereal vehicles (cf. Henry 1986). John Locke suggested that God could, if he wished, imbue matter with the property of thought.

In addition to these quite specific Christian doctrines there was a range of more general theological ideas that needed to be rethought: final causes (and causation in general), divine action, miracles, free will and determinism.

Causation, divine action, and laws of nature

The demise of the Aristotle's four causes and the replacement of this system with a mechanical and atomistic conception of matter entailed a radical reconceptualisation of causation in nature and of how God acts in the world. Because the world was no longer populated with discrete kinds of things, each with their own causal powers, but rather with objects made up of different combinations of essentially inert particles, the causal efficacy of things had to be located elsewhere. For many thinkers, the direct source of causation in the world was God. (In an interesting parallel, this view of causation, known as occasionalism, had also been adopted by Arabic atomists in the early Middle Ages.) Because God always acts in a lawful fashion, the mathematical regularities that we observe in the world are the direct consequence of God's imposition of his will on the invisible and inert particles that constitute the basic stuff of the world. For Aristotelians, natural changes take place because of active powers in nature, and God concurs in these operations. After Descartes, natural philosophers tended to think rather that natural change is brought about directly by God (see Des Chene 2000, Nadler 1998, Garber 1987). The regularities which Aristotelians located in the discrete natures of things are, in the post-Cartesian world, attributed to the lawful volitions of the Deity. (On the origin of the idea of laws of nature, see Henry 2004, Harrison 2008, Harrison 1995.)

Consider, for example, Descartes' explanation of motion: 'God imparted various motions to the parts of matter when he first created them, and he now preserves all this matter in the same way, and by the same process by which he originally created it' (Descartes 1984, vol. 1 p. 240: cf. vol. 1 p. 243 and vol. 2 p. 33). God, on this account, is not only the first unmoved mover, but remains the only source of motion in the universe. Motion, thus understood, is God's successive creation of an object in adjacent spatial locations.

While the philosophy of Descartes is often supposed to stand in stark contrast to that developed by the English experimentalists, in many respects it provided the metaphysical foundations for their experimental natural philosophy. Robert Boyle, for example, reiterates the basic Cartesian position that 'the laws of motion ... did not necessarily spring from the nature of matter, but depended on the will of the divine author of things' (Boyle 1966, vol. 5 p. 521). Isaac Barrow, the first Lucasian Professor of Mathematics, and Newton's immediate predecessor in that Chair, endorsed the Cartesian view of causation, writing that 'the efficient Cause of all Things is God' (Barrow 1734, lecture VII p. 109).

Even Newton himself, despite his rejection of the Cartesian world system, had much more in common with Descartes than with the Aristotelians. In the Preface to the first edition of the *Principia*, he highlights the difference between his own approach and that of 'the ancients', pointing out that while the ancients (and, by implication, their modern followers) had investigated nature by seeking the inherent forms and qualities of things, the moderns had sought 'to reduce the phenomena of nature to mathematical laws' (Newton 1999, p. 381). Newton's disciples rehearsed this message. Richard Bentley, the first Boyle lecturer, wrote that 'God uses no other means, instruments or applications in these productions [of nature], than his bare word or command' (Bentley 1838, p. 75). Samuel Clarke, who argued Newton's case in the dispute with Leibniz, contended that 'the *Course of Nature*, cannot possibly be any thing else, but the *Arbitrary Will and pleasure of God* exerting itself and acting upon Matter continually' (Clarke 1738, vol. II p. 698).

Later, Newton was to speculate about what all of this might mean for gravitational attraction, which he insisted was not a 'quality' inhering 'in the specifick forms of Things' but rather a phenomenon that arose from 'general Laws of Nature' (Newton 1952, p. 401). Again, his disciples preached the same message. Isaac Barrow confidently declared that 'it may be proved, in its due place, that this gravity, the great basis of all mechanism, is not itself mechanical, but the immediate *fiat* and finger of God, and the execution of divine law' (Barrow 1885, vol. II p. 303). Gravity thus presented a specific example of the more general claim that the regularities of nature were explicable in terms of the imposition of God's will directly on matter, rather than as arising out of powers embedded in natural things.

The general tendency toward corpuscularian matter theory and mechanical explanation might be regarded as typical of the seventeenth century, partly because from the vantage point of twenty-first century these ideas seem to be on the 'right track'. But there were other options. As already indicated, the German philosopher Leibniz was opposed to much of the mechanical philosophy and its accompanying matter theory. The mechanical philosophy, he believed, failed the test of the Principle of Sufficient Reason. Moreover, there seems to be no plausible theory of inelastic collisions of basic particles. Leibniz developed a monist metaphysics as an alternative to the mechanical philosophy.

An interesting compromise position was advocated by the Cambridge Platonists. These individuals subscribed to the corpuscular hypothesis, but were concerned by what they regarded as its incipient materialism and deism. Accepting the inertness of matter, they suggested that some non-material spirit must act as an intermediary between God and the world. Henry More enlisted what he called 'the Spirit of Nature' or an 'hylarchic principle' to fulfil this function. More attributes the lawlike regularities of nature to this principle, which is said to embody 'certain general Modes and Lawes of Nature' (More 1662 Preface, p. xvi). One intriguing aspect of his thought is the conviction that spirit can be extended. Hence, the spirit of nature is extended through the cosmos, the human soul is extended throughout the body and, indeed, space offers an example of infinitely extended, yet non-material, medium. (Later, in what seems to be a direct parallel, Newton was to propose infinite space as God's sensorium: Newton 1952, p. 402.) Henry More's fellow Platonist Ralph Cudworth spoke similarly of a 'plastick nature', which also acted as God's intermediary in the natural world. Again, the laws of nature and divine purposes are embodied in this immaterial medium (Hutton 2001). Both Cudworth and More sought to combine Cartesian corpuscularianism with Platonism. To some extent, their ideas are also redolent of the old Platonic idea of a world soul, or the Stoic idea of seminal principles (Clericuzio 2001). (It is noteworthy that Boyle also spoke explicitly of seminal principles, although he later seems to have discarded the concept (Anstey 2002); and that Descartes toyed with the idea of 'seed-like' principles being operative at the creation (Harrison 2000).)

The criticisms and modifications of the corpuscular philosophy often turned on theological questions, and highlight what was at stake from a religious perspective – in particular, issues relating to materialism, determinism, occasionalism, and voluntarism. For advocates of the corpuscular philosophy, however, none of these threw up insurmountable difficulties, and arguments were developed to show that the new philosophy was far more compatible with Christian theology than the Aristotelianism that preceded it. Indeed, such is the enthusiasm for a scientifically informed theology at this time that it possible to speak of the emergence of a new 'secular theology' promulgated not by theologians, but by natural philosophers (Funkenstein 1986, pp. 3-9). This new kind of theology, moreover, played a key role in providing social legitimation for the new natural philosophy. Modern science, on this view, is forged in collaboration with theology, rather than in opposition to it (see especially Gaukroger 2006).

In drawing attention to these features of early modern science, my aim is not to necessarily to provide a celebratory account of this close relation between science and religion at this time, but to spell out some of its possible implications for the relations between science and religion that were to follow.

Matter and meaning

Up to this point we have been mostly concerned with competing theories of matter, and the replacement of one such theory with another. However, the theme of this book concerns the questions of matter *and meaning*, and I want to devote a few paragraphs to the implications of the new matter theory of the seventeenth century for the question of the meaning of nature. My suggestion, in brief, is that the new philosophy was accompanied by an evacuation of theological meaning from the natural world. I do not mean by this that the new matter theory was somehow less theologically informed than the matter theory of the Middle Ages. If anything, the reverse was the case. But the question of the theological *meanings* of the natural world in the Middle Ages was not located in discussions of matter theory. Rather, the notion that there were symbolic meanings in nature was nourished by an emblematic hermeneutics of nature (cf. Harrison 1998). When we consider the medieval system of allegorical interpretation, the descriptions of the symbolic meanings of animals and plant that we encounter in the medieval bestiaries, and the deeply influential notion of the 'book of nature', it becomes clear that, running parallel to the Aristotelian natural philosophy, there is a symbolic ordering of nature. We might thus speak of two distinct ways of locating order in the natural world during the Middle Ages, one to do with causal relations (Aristotelian) and another to do networks of meaning (Christian Neoplatonic).

Integral to the emblematic understandings of nature is the conviction that God has created visible things to act as signs that signify hidden theological truths. This view is partly indebted to the basic distinction with which we first began – that between the appearance of things, and the invisible reality that lies behind them. More important was the theological version of this view offered by St Paul: 'the invisible things of him from the creation of the world are clearly seen, being understood by the things that are made.' (Rom. 1:20, KJV). Accordingly, the primary locus of the theological significance of nature in the Middle Ages does not lie in the relatively 'thin' derivation of God's existence from principles of Aristotelian natural philosophy, as encountered, for example, in Aquinas's 'five ways'. (Given that this natural philosophy was developed in complete independence from Christian theology, this is hardly surprising.) Rather, the theological significance of nature is vested primarily in its symbolic meanings. This view of nature was intimately associated with the practice of the allegorical interpretation of scripture. Not only was God said to have authored 'two books', but the interpretation of these two books, in the Middle Ages, was part of an integrated hermeneutical practice. This changes radically in the seventeenth century, partly because of the demise of allegorical interpretation.

One way to see the significance of this idea is to compare 'book of nature' metaphors in the Middle Ages with those of the early modern period. In the twelfth century we find Hugh of St Victor declaring that 'the whole sensible world is like a kind of book written by the finger of God – that is, created by divine power – and each particular creature is somewhat like a figure, not invented by human decision, but instituted by the divine will to manifest the invisible things of God's wisdom' (Hugh of St Victor, 814B). Bonaventure writes in the thirteenth century that the creature of the world is like a book, in which the divine Trinity is reflected (Bonaventura 1963, p. 104). Book of nature metaphors of the seventeenth century, by way of contrast, speak not of a symbolic ordering of nature in terms of surface resemblances and theological meanings, but instead of a mathematical ordering, or of mechanical structures. Galileo thus asserted that the book of nature was written in mathematical language, and that one needed to be a mathematician to read it (Galileo 1957, pp. 237f). In this new understanding, the book of nature does not convey theological truths symbolically, but instead provides premises upon which arguments from natural theology can be constructed. The transition from the Middle Ages to the early modern period thus sees a quite dramatic transformation in the way in which the theological significance of nature is understood. In the Middle Ages, a *theology of nature* enables nature to be interpreted in such as way as to convey a wide range of theological truths – including the triune nature of God. In the early modern period, a new *natural theology* makes possible relatively restricted theological inferences about the divine nature. From the remarkable mechanisms evident in nature, we infer God's power and wisdom (Harrison 1998, pp. 161f: Harrison 2006). It is no exaggeration to say that that this approach makes possible the birth of a new natural theology.

The implications of the new matter theory were thus quite far-reaching. In conclusion, I want to point to some other long term consequences of corpuscular philosophy for the relationship between science and religion.

Conclusion: the legacy of the corpuscular hypothesis

Perhaps the first point to make by way of a protracted conclusion is to highlight the fact that the views of God's intimate relation to the world outlined by the seventeenth century natural philosophers proved almost immediately to be susceptible to non-theistic interpretations. It is a small step from the claim that God is the direct cause of everything to the assumption that he is the cause of nothing. Newton had sought to protect his system from this interpretation by insisting that God was not only responsible for the constant upholding of regular laws, but that on occasion he was also needed to iron out the instabilities of planetary orbits. However, it was not long before Pierre-Simon Laplace (1749-1827) could famously declare, of Newton's intervening Deity, that 'he had no need of that hypothesis'. This was perhaps the first failure of the 'God of the gaps' principle. In retrospect, it was also one of the most remarkable episodes in the history of science, for the Newtonian system, in which God is ever-present and integral to the workings of the system, was almost completely naturalised just a few decades after the death of its author (Hahn 1981). It is also significant that at this time Laplace and others began circulating a story that Newton had suffered a breakdown in the 1690s. A deranged Newton, the story went, had then turned to theology. This myth, which effectively quarantined Newton's science from his theological commitments, made him an appropriately sanitized exemplar of Enlightenment ideals (Snobelen 2004: Manuel 1963, p. 5). This naturalization effectively returns gravity to the status of an 'occult quality' – the much derided scholastic category which proposed as an explanation what was in effect a mere labelling exercise – and undermines the original theological foundation of the notion of laws of nature. The implication of this move is far-reaching. As philosopher of science Nancy Cartwright has recently put it: 'no God, no laws' (Cartwright 2005).

A second point relates to the idea of God as a craftsman who directly creates things in the forms in which we encounter them in nature. The conviction that the created order was mechanical in its operations naturally suggests this model of divine activity. With the demise of a symbolic ordering of nature, the predominant analogy of God's relation to the world changes from that of author to book, to that of watchmaker to watch. If William Paley made this latter analogy famous, it had already been explicitly set out in seventeenth-century understandings of the theological implications of the corpuscular theory and the mechanical philosophy. Such an image contrasts not only with symbolist understandings (nature as a book of symbols), but also with organic models (nature as a self-organising creature), and, in relation to the latter, particularly with Augustine's suggestion (although this was not widely adopted) that God's creative ends might have been accomplished as the consequence of a long developmental process set in train at the beginning of the creation. (Admittedly, in his *Principles of Philosophy*, Descartes had proposed a kind of cosmic evolution, but this had been proposed as a hypothetical model and was susceptible to accusations of Epicureanism.) As we know, the image of the designing Deity fared poorly in the wake of Darwinian ideas of evolution by natural selection.

Thirdly, although seventeenth century matter theory, along with its deterministic implications, was significantly modified in the twentieth century in the wake of quantum mechanics, nonetheless it continues to have a major influence on our thinking. One legacy is an ongoing difficulty in speaking about divine action in the natural world. Since the seventeenth century, this tends to be an all-or-nothing affair, respectively mirroring the approaches of occasionalists such as Malebranche or naturalists such as Laplace. Another lingering influence is a conceptual confusion about the status of laws of nature: do we discover these in nature, or is some kind of lawful intelligibility assumed in our investigations of nature? Most seriously, perhaps, atomic matter theory and the seductive idea of mechanistic explanation have so influenced our understanding of science that the reductionist approach that they represent is often assumed to be essential to, or in some way constitutive of, genuinely scientific endeavour. The idea that all phenomena are to be explained in terms of what is happening at the most fundamental level of reality has long left the confines of matter theory, and has been elevated to the status of a metaphysical doctrine. Atomism, understood thus as a powerful heuristic, is part of the mixed legacy of early modern matter theory. Arguably, the assumption that all phenomena are susceptible to reductionist analysis of this kind is not well founded.

Finally, to end on a more positive note, in light of the theological motivations of the relevant historical actors, and the fact that many of their innovations were grounded in theological presuppositions, the common view that modern science was forged largely independently of theological considerations, or indeed in opposition to them, fails the most basic test of historical veracity. And if there is any truth to the considerations set out above, the theological origin of these ideas has left an indelible mark on our present scientific understandings.

Bibliography

Anstey, P. 2002. 'Boyle on Seminal Principles', *Studies in History and Philosophy of Science, Part C: Biological and Biomedical Sciences*, 33 pp. 597-630.

Aristotle, 1933. *Metaphysics*, tr. H. Tredennick, 2 vols. London: Heinemann.

Aquinas, T. 1932. *Quaestiones disputatae de potentia dei*. English translation, *On the Power of God*, tr. English Dominican Fathers. London: Burns, Oates & Washbourne.

—. 1934. *Summa contra gentiles*. tr. English Dominican Fathers, 5 vols. London: Burns, Oates & Washbourne.

—. 1961. *Commentary on the Metaphysics of Aristotle*, tr. John P. Rowan, 2 vols. Chicago: Henry Regnery.

Armogathe, J.-R. 1977. *Theologia cartesiana: L'explication physique de l'Eucharistie chez Descartes et Dom Desgabets*. The Hague: Nijhoff.

Augustine 2002. *De Genesi ad litteram* in *Works of Saint Augustine*, I, vol. 13, tr. Edmund Hill and Matthew O'Connell. New York: New City Press.

Barrow, I. 1734. *The Usefulness of mathematical learning explained and demonstrated*, tr. John Kirby. London.

—. 1885. 'Maker of heaven and earth' (Sermon XII), in *Theological works*. 3 vols. London.

Bentley, R. 1838. *The Works of Richard Bentley, D.D.*, ed. Alexander Dyce. Vol III, Theological Writings. London: Macpherson.

Bonaventura 1963. *The Breviloquium* [1257], tr. J. de Vinck. Paterson, NJ: St Anthony Guild Press.

Bourg, J. 2001. 'The Rhetoric of Modal Equivocacy in Cartesian Transubstantiation', *Journal of the History of Ideas* 62 pp. 121-140.

Boyle, R. 1966. *The Christian Virtuoso*, in The Works of the Honourable Robert Boyle, ed. Thomas Birch, 6 vols. [1772]. Hildesheim.

Cartwright, N. 2005. 'No God; No Laws', *Dio, la Natura e la Legge. God and the Laws of Nature*. Angelicum-Mondo X, pp. 183-190.

Clarke, S. 1738. 'The Evidences of natural and revealed religion', *The works of Samuel Clarke, D.D.*. 2 vols., London.

Clericuzio, A. A. 1990. 'A redefinition of Boyle's Chemistry and Corpuscular Philosophy', *Annals of Science* 47 pp. 561-589.

—. 2001. 'Gassendi, Charleton and Boyle on Matter and Motion', in Lüthy, Murdoch and Newman (eds.), *Matter Theories*.

Descartes, R. 1984. *The Philosophical Writings of Descartes*, tr. John Cottingham, Robert Stoothoff and Dugald Murdoch (2 vols.), Cambridge: Cambridge University Press.

Des Chene, D. 2000. 'On Laws and Ends: A Response to Hattab and Menn', *Perspectives on Science,* 8 pp. 144-63.

Funkenstein, A. 1986. *Theology and the Scientific Imagination*. Princeton: Princeton University Press.

Galileo, G. 1957. *The Assayer,* in *Discoveries and Opinions of Galileo* tr. Stillman Drake. New York: Anchor.

Garber, D. 1987. 'How God causes motion: Descartes, divine substance, and occasionalism', *Journal of philosophy,* 84 pp. 567-80.

Gaukroger, S. 2006. *The Emergence of a Scientific Culture*. Oxford: Clarendon Press.

Gilson, E. 1961. *The Christian Philosophy of Saint Augustine*. London: Victor Gollancz.

Glasner, R. 2001. 'Ibn Rushd's Theory of *Minima naturalia*', *Arabic Science and Philosophy* 11 pp. 9-26.

Hahn, R. 1981. 'Laplace and the vanishing role of God in the physical universe', in Harry Woolf (ed), *The analytic spirit: essays in the history of science in honor of Henry Guerlac*. Ithaca: Cornell University Press.

Harrison, P. 1995. 'Newtonian Science, Miracles, and the Laws of Nature', *Journal of the History of Ideas*, 56 pp. 531-53.

—. 1998. *The Bible, Protestantism and the Rise of Natural Science* Cambridge: Cambridge University Press.

—. 2000. 'The Influence of Cartesian Cosmology in England', in S. Gaukroger, J. Schuster, and J. Sutton (eds.), *Descartes' Natural Philosophy*. London: Routledge.

—. 2006. '"The Book of Nature" and Early Modern Science', in K. van Berkel and Arjo Vanderjagt (eds.), *The Book of Nature in Early Modern and Modern History*. Leuven.

—. 2008. 'The Development of the Concept of Laws of Nature', in Fraser Watts (ed.), *Creation: Law and Probability*, pp. 13-36. Aldershot: Ashgate.

Henry, J. 1986. 'A Cambridge Platonist's Materialism: Henry More and the Concept of Soul', *Journal of the Warburg and Courtauld Institutes* 49, pp. 172-95.

—. 2004. 'Metaphysics and the Origins of Modern Science: Descartes and the Importance of Laws of Nature', *Early Science and Medicine,* 9 pp. 73-114.

Hugh of St Victor, *De tribus diebus*, Migne PL XXVI, 814B.

Hutton, S. 2001. 'Ralph Cudworth, God, Mind and Nature', in Robert Crocker (ed.), *Religion, Reason and Nature in Early Modern Europe*. Dordrecht: Springer.

Lüthy, C., Murdoch, J. E. and Newman, W. R. (eds.) 2001. *Late Medieval and Early Modern Corpuscular Matter Theories*. Leiden: Brill.

Mahony, M. 1998. 'The Mathematical Realm of Nature', in Garber and Ayers (eds.), *Cambridge History of Seventeenth Century Philosophy*, vol. 1, pp. 702-55.

Manuel, F. 1963. *Isaac Newton, historian*. Cambridge, MA: The Belknap Press of Harvard University Press.

More, H. 1662. *A Collection of Several Philosophical Writings*. London.

Nadler, S. 1988. 'Arnauld, Descartes, and Transubstantiation: Reconciling Cartesian Metaphysics and Real Presence', *Journal of the History of Ideas* 49 pp. 229-46.

—. 1998. 'Doctrines of explanation in late scholasticism and in the Mechanical philosophy', in Garber and Ayers (eds.), *Cambridge history of seventeenth-century philosophy*, vol. 1, pp. 513-52.

Newman, W. R. 2006. *Atoms and Alchemy: Chymistry and the Experimental Origins of the Scientific Revolution*. Chicago: University of Chicago Press.

Newton, I. 1952. *Opticks* [4th edn.] New York: Dover.

—. 1999. *The Principia. Mathematical Principles of Natural Philosophy*, tr. I. Bernard Cohen and Anne Whitman (Berkeley: University of California Press).

Palmieri, M. 1997. *Civil Life II*, in Jill Kraye (ed.), *Renaissance Philosophical Texts*, Vol. 2 pp. 149-172. Cambridge: Cambridge University Press.
Snobelen, S. 2004. 'To Discourse of God: Isaac Newton's Heterodox Theology and his Natural Philosophy', in Paul Wood (ed.), *Science and Dissent in England, 1688-1945*, pp. 39-66. Aldershot: Ashgate.
Wilson, C. 2008. *Epicureanism at the Origins of Modernity*. Oxford: Oxford University Press.
Wolfson, H. A. 1976. *The Philosophy of the Kalam*. Cambridge, MA: Harvard University Press.

CHAPTER FIVE

THEOLOGY AND THE MEANING OF MATTER IN THE EARLY MODERN PERIOD: A RESPONSE TO PETER HARRISON

JOHN HENRY,
UNIVERSITY OF EDINBURGH

Professor Harrison has gone far beyond the claims of his title and has offered us a succinct but evocative history of the interactions between theology and matter theory from the Ancient Greeks to Pierre-Simon Laplace (1749–1827) and beyond. Along the way he has wonderfully evoked the perpetual tensions between materialism and immaterialism in different theologies, and the implications of these for developments in physics, such as new theories of causation and new conceptions of the laws of nature. I cannot fault Professor Harrison's account: the only complaint I can make is testimony to how well he has fulfilled his brief. That complaint is that he could have told us more; but since it is a golden rule to always leave your audience wanting more, it seems that once again he has performed ideally!

Since I see no need to take issue with any of the points Harrison makes, perhaps I can be allowed to add to his account by considering one aspect of the story in more detail. My aim is not to point to a lacuna in Harrison's account, but rather to reassure readers that the close relationship between science and religion which he has illustrated is not merely the result of his wide-ranging approach. It might be suspected by those who prefer to believe that science and religion represent completely different and incompatible world-views that the fertile interactions between them which Harrison has outlined are nothing more than illusions created by an impressionistic, broad-brush, style of history. I hope to provide a corrective to such a view, by showing that even a detailed scrutiny of the role of matter theory in history can be shown to depend upon close interactions between theologians on the one hand, and natural philosophers on the other.

Bearing in mind our main theme, matter and meaning, it seems to me that Harrison's idea that matter was *evacuated* of its meaning in the seventeenth century is particularly insightful (see p. 49 above), and I will try to show that those insights remain even as we descend into detail. One important aspect of the case history that I discuss is that it demonstrates that the evacuation of meaning from matter was not an exclusively scientific endeavour. Again, those who wish to separate science from religion may suppose that the attempt to remove various significances with which matter had become invested was a 'hard-headed', rationalist, secularist, and (therefore) scientific enterprise. Those who think this way might also suppose that Churchmen, by contrast, would always prefer to see matter infused with various mystical and spiritual properties, and meanings. The historical reality reveals that such notions are merely the wishful thinking of those who want to maintain the surely false idea that science and religion are incompatible and entirely distinct worldviews.

There is a famous episode in the history of matter theory at the end of the seventeenth century, when the philosopher John Locke (1632–1704) suggested that matter, in its own right, might be capable of thinking (this is mentioned briefly by Harrison (see p. 46 above): for a major historical study of this notion, see Yolton 1984). For Locke's contemporaries, this was immediately seen as undermining the prevailing Cartesian distinction between *res extensae* – extended things – and *res cogitantes* – thinking things. For René Descartes (1596–1650), thinking things were non-physical things: not only were they incorporeal, but they could not even be said to occupy, or to exist in, space – they were somehow completely separate from the physical world. By complete and non-overlapping contrast, *res extensae*, or material things, were constitutionally incapable of thinking (Rozemond 2002).

Now it might seem that, by insisting that matter might be capable of thinking, Locke was revealing himself to be an out and out materialist, rejecting the dualism between body and soul, which was not only Cartesian but also Christian. (The Roman Catholic Church had proclaimed the soul to be immortal by virtue of its nature at the Lateran Council in 1513, and even went on to insist that all philosophers should defend this view by proving the soul's immortality rationally. Descartes may be seen to have been responding to this call: see Schroeder 1937, pp. 483-7, and Kessler 1988, p. 495.) But this conclusion would be wrong. Locke did not object to Cartesian dualism on *philosophical* grounds, but on *theological* grounds. His concern was not to promote materialism, but to preserve the omnipotence of God. 'The question is,' Locke wrote, 'whether God can, if he please, bestow on any parcel of matter, ordered as he thinks fit, a faculty of perception and thinking' (Locke 1892, vol. ii p. 398). For Locke, the answer had to be, 'yes, he can.'

Cartesian dualism, which at that time had been taken up by many theologians to bolster their own Christian dualism, implied that matter was categorically incapable of thinking, and that, therefore, not even God could make matter think (see Henry 1989). It was this which stimulated Locke's claim about thinking matter. If God chose to make matter think, Locke insisted, nothing Descartes can say is going to prevent this from coming about. As a matter of fact Locke did not believe that matter could think. Like everyone else at the time, he subscribed to the view that what thinks in us is an immaterial something, but he may have wanted to emphasise God's omnipotence in such matters as an indirect way of defending the natural philosophy of his friend Isaac Newton (1642–1727).

Newton, after all, was at this time proposing, and defending, the idea that matter could act on other matter, at a distance, across vast distances of empty space (this was the implication of the universal principle of gravity). Now action at a distance had always been regarded as an impossibility: a thing cannot act where it is not (Hesse 1961). Instinctive feelings about the impossibility of action at a distance are so strong that many scholars have refused to believe that Newton accepted them, in spite of his own clear pronouncements that he did (Henry 1994): Newton implicitly assumed such action in his *Principia mathematica* (1687), and explicitly defended it in the 'Queries' which he added to his *Opticks* in 1704 (Newton 1999, p. 382: Newton 1952, Queries 1, 4, 20, 21, 29 and 31). It seems perfectly clear that Newton was able to accept this notion without qualms because of his faith in the omnipotence of God. If God chooses to make matter act where it is not, by a force of gravitational attraction, nothing can prevent it. Certainly, this seems to be implicit in his famous comment in the *Opticks*:

> God is able to create Particles of Matter of several Sizes and Figures, and in several Proportions to Space, and perhaps of different Densities and Forces, and thereby to vary the Laws of Nature, and make Worlds of several sorts in several Parts of the Universe. At least, I see nothing of Contradiction in all this (Newton 1952, pp. 403-4).

Indeed, there are second hand reports from the 1690s that Newton believed that God was directly responsible for gravitational attraction – that is to say, that God directly moved all falling bodies. This is a position similar to the occasionalism mentioned by Harrison (p. 46 above: on occasionalism, see, for example, Schmaltz (2008)). Those who attributed this view to Newton were said to smile at this idea, but I don't think they were smiling in a good way: more likely it was with an air of superiority. But it seems to me that we shouldn't trust these reports. These smiling men were perhaps incapable of accepting the notion that matter could be made to act at a distance, in the same way that Locke's critics could not accept that matter could be made to think. Noting Newton's emphasis upon the role of God in his discussions of how matter can act at a distance, these men perhaps assumed that he was invoking God's direct intervention. All the evidence from Newton's own pronouncements, however, is that he believed gravity to be an attribute of matter, a secondary cause, which God had endowed upon matter at the Creation (cf. Westfall 1980, pp. 509-10: Henry 1994).

Here then, in Newton and in Locke, we have a depiction of matter which, far from being the passive inert matter of the Cartesian mechanical philosophy, is intrinsically active (whether attracting other matter at a distance, or thinking), and thereby invested with theological meaning. It is interesting to note, I think, that this view of matter became immensely influential throughout the eighteenth century. Consider, for example, the comment of the Swiss mathematician, Daniel Bernoulli (1700–1782), to Leonard Euler (1707–1783) in a letter of February 1744:

> I cannot hide from you that on this point I am a complete Newtonian, and I marvel you so long adhere to the Cartesian principles;... if God could create a soul whose nature is incomprehensible to us, so could he impress a universal attraction on matter, even if such an attraction is beyond our comprehension (quoted in Wilson 1992, p. 399).

One way of summing this up is to say that Newton and his followers wanted a theory of matter that was undeniably theological in its meaning: it could not be understood without assuming the existence of God. As Newton himself said of one of his early speculations about the nature of matter: 'we cannot postulate bodies of this kind without at the same time supposing that God exists, and has created bodies in empty space out of nothing' (Newton 1962). For Descartes and his followers, by contrast, matter was abandoned as something with no theological meaning, and all such meaning was invested in the immaterial, the spiritual – in short, the other side of the dualist coin.

It was only in the nineteenth century that scientists began to turn against the Newtonian theologically-dependent view of matter. It was in the nineteenth century that non-Euclidian geometries enabled physicists to conceive of curvatures in space, and these curvatures could be invoked to explain gravitational movements without invoking actions at a distance between bits of matter: the Earth orbited the Sun, for example, because it was caught in a dip in space caused by the mass of the Sun (van Lunteren 1988: Gray 2006). Matter was once more inactive and, as Harrison has suggested, stripped of theological meaning.

In nineteenth-century physics, therefore, we can see a clear attempt by some secularising scientists to strip matter of the theological meaning that it had long since held in the Newtonian tradition. In particular, the mathematician William Kingdon Clifford (1845–1879) developed ideas about the curvature of space as a deliberate attempt to deprive Peter Guthrie Tait (1831–1901), professor of natural philosophy at Edinburgh University, of the aether theory which he had used to prove the existence of God and the immortal soul in his *Unseen Universe* (Stewart and Tait 1876: Clifford 1885. See also Smith and Wise 1989: Wilson 1991). It would be a mistake, however, to suppose that it was always secularising scientists who sought to strip matter of meaning. It is important to note that the process was never one-sided, and that theologians, as much as scientists, were responsible for stripping matter of its meaning.

We can see this, for example, in early modern efforts to extirpate not only witchcraft, but even popular superstitions. Demonologists had always made a distinction between what could be achieved by so-called natural magic, and what could not. A cunning man or woman might be able to make a cow go dry, or to make chickens cease from laying eggs by entirely natural means – in which case, though they might be guilty of a crime against a neighbour, there was no suggestion that demonic aid was involved. It was not possible, however, for a human being to fly. If, therefore, a cunning man or woman claimed that they had flown, or had made somebody else fly, they must have been deceived, and so had been (perhaps unwittingly) involved with the Devil or some lesser demon. In such cases, something like an ointment, or a charm, or a spell, which was supposed to make the person capable of flying, could not be naturally efficacious, and so was deemed by the demonologists to be merely a *sign* of compact with demons (see Clark 1997 for a study of the theories of demonology).

The background to this is a pre-Cartesian matter theory, either based in Scholastic Aristotelianism, or in a debased version of this which had trickled down into popular culture, in which bodies had various powers or qualities by which they could affect other things. Broadly speaking, within these traditions, matter could be naturally efficacious in a number of ways, even though some of these ways were mysterious or occult (for fuller discussions, see Henry 1986 and Henry 2008). This theory of matter provided too many opportunities for people to insist that they took no part in sorcery, had no commerce with demons, but had merely exploited the natural powers of matter. It was extremely important for the Churches, however, to be able to distinguish between what had been accomplished by natural means, and what could only have been accomplished with demonic aid (or more usually in cases of the latter, what could only be claimed to have been accomplished as a result of demonic deception – so if a man genuinely claimed he had walked through a wall to steal something, he might not actually be guilty of theft, but he must have allowed himself to be deceived by a demon). Churchmen were very keen, therefore, to understand the natural powers of matter in order to be able to distinguish between what could be accomplished by exploiting the natural powers of things, and what could not (Clark 1984: Clark 1997).

But even far below the level of possible sorcery, in the realm of superstition, too many (according to the Churches, anyway) believed in the magical efficaciousness of ordinary everyday objects. Superstition, of course, was seen as a threat to sound religion, undermining the efforts of the Church by offering 'vain observations' as a substitute for proper religious observance. Such vain observations often involved attributing some power or efficacy to bodies beyond those powers known to sound natural philosophy. In the fight against superstition, therefore, Churchmen adopted similar attitudes to the demonologists: seeking to distinguish between the real properties of things and those merely attributed without foundation, belief in the latter being seen as a clear sign of a lack of true faith (Clark 1991).

It seems clear that seventeenth-century Churchmen, concerned by the flourishing of superstition, saw Cartesianism, with its extremely austere view of what matter was capable of, as a godsend. There was now no need to try to educate the common people in the niceties of Aristotelian matter theory to enable them to distinguish between the natural efficacy of matter and a bogus efficacy merely attributed to it. In a Cartesian world, it was possible to insist that matter is simply a dead thing, so that no material body has any significant efficacy (except by impact in collision). Churchmen wanted to see the disenchantment of the material world just as much, if not more so, than mechanical philosophers such as Descartes, or Robert Boyle (1627–1691).

The aim, therefore, as Harrison has said, was to strip matter of its meaning. It seems clear that in doing so, the theologians wished to force a renewed focus on the alternative to matter – the incorporeal, or spiritual – as the source of all activity and power. This is most clearly seen in the so-called Cambridge Platonists, Henry More (1614-1687) and Ralph

Cudworth (1617-1688), both of whom insisted that any activity in the world, even the falling of a stone, must be attributable to the causative powers of an active spirit, because matter is completely passive and incapable in itself of any activity whatsoever (cf. Henry 1990: Henry 2007). But we know that, for whatever reasons, atheism had a momentum of its own by the late seventeenth century, and was capable of turning any philosophy to its advantage (see Buckley 1987). Descartes, like other philosophers before him, had failed to give a cogent account of how his completely non-physical *res cogitantes* (which, remember, could not even be said to exist in physical space) could interact with the physical world. And yet, his physics seemed perfectly intelligible, and to work perfectly well, without such an account. It was an easy step, therefore, for the irreligious to dispense with the incorporeal all together. A Cartesian physics of inert matter was sufficient unto itself. The disenchanted world of dead matter, in spite of the rearguard efforts of even so great a physicist as Isaac Newton, became all that was required, and the world itself, not just matter, began to be stripped of theological meaning.

Here, then, we see the historical background to the legacy mentioned by Harrison: a legacy that makes God's action in the world an all-or-nothing affair (see p. 52 above). For the atheists, represented by Laplace, God's action in the world is meaningless. For the faithful, however, in a world in which they long ago colluded with the view that matter in itself has no theological meaning, they have to suppose that God must intervene directly.

Having said that, however, it is well known that in the ranks of scientists there are more believers among physicists than there are among biologists. Perhaps, in the world of sub-atomic physics, which seems perpetually to be teetering on the brink of meaninglessness, physicists are looking anew for God (and meaning) as well as for the Higgs boson. Indeed the Higgs boson, which is supposed to account for mass in the strange world of subatomic physics, has been dubbed the 'God particle' by popular science writers (cf. Achenbach 2008). Perhaps the time is ripe for scientists and theologians to once again re-assert the theological meaning of matter.

They will have to be careful, however, for as we all know history has a habit of repeating itself. Just now there is a huge proliferation of so-called New Age belief systems, and many of these make claims about, or simply take for granted, various magical and mystical properties of matter. These ideas about matter and its properties are further from received scientific views than the ideas of early modern cunning men and cunning women were from the learned views that were current in their time. Furthermore, popular writers, such as Fritjof Capra, author of *The Tao of Physics* (1975), find it easy to exploit contemporary collective ignorance and claim that there are congruences between theories in modern physics and assumptions about the nature of the physical world in eastern mystical religions (Capra 1975: for a devastating critique, see Restivo 1985). If we are ever to understand the real meaning of matter, it is surely important that scientists and theologians pay very careful attention to one another – and ideally, as they used to in the early modern period, think like one another too.

Bibliography

Achenbach, J. 2008. 'At the Heart of All Matter: The Hunt for the God Particle', *National Geographic Magazine*, March 2008.

Buckley, M. J. 1987. *At the Origins of Modern Atheism*. New Haven: Yale University Press.

Capra, F. 1975. *The Tao of Physics: An Exploration of the Parallels between Modern Physics and Eastern Mysticism*. London: Wildwood House.

Clark, S. 1984. 'The Scientific Status of Demonology' in Brian Vickers (ed.), *Occult and Scientific Mentalities in the Renaissance*, pp. 351-74. Cambridge: Cambridge University Press.

—. 1991. 'The Rational Witchfinder: Conscience, Demonological Naturalism and Popular Superstitions', in Stephen Pumfrey, Paolo Rossi, and Maurice Slawinski (eds), *Science, Culture and Popular Belief in Renaissance Europe*, pp. 222-48. Manchester: Manchester University Press.

—. 1997. *Thinking with Demons: The Idea of Witchcraft in Early Modern Europe*. Oxford: Clarendon Press.

Clifford, W. K. 1885. *The Common Sense of the Exact Sciences*, edited by Karl Pearson. London: K. Paul, Trench.

Gray, J. 2006. *Worlds Out of Nothing: A Course in the History of Geometry in the 19th Century*. Dordrecht: Springer.

Henry, J. 1986. 'Occult Qualities and the Experimental Philosophy: Active Principles in pre-Newtonian Matter Theory', *History of Science*, 24, pp. 335-81.

—. 1989. 'The Matter of Souls: Medical Theory and Theology in Seventeenth-Century England', in R. K. French and A. Wear (eds.), *The Medical Revolution in the Seventeenth Century*, pp. 87-113. Cambridge: Cambridge University Press.

—. 1990. 'Henry More versus Robert Boyle: The Spirit of Nature and the nature of Providence', in Sarah Hutton (ed.), *Henry More (1614-1687): Tercentenary studies*, pp. 55-75. Kluwer Academic Publishers, Dordrecht.

—. 1994. '"Pray do not ascribe that notion to me": God and Newton's Gravity', in James E. Force and Richard H. Popkin (eds), *The Books of Nature and Scripture: Recent Essays on Natural Philosophy, Theology and Biblical Criticism in the Netherlands of Spinoza's Time and the British Isles of Newton's Time*, pp. 123-47. Dordrecht: Kluwer Academic Publishers.

—. 2007. 'Henry More (1614-1687)', *Stanford Encyclopaedia of Philosophy* (August, 2007): http://plato.stanford.edu/entries/henry-more/ (accessed 3 November 2009).

—. 2008. 'The Fragmentation of the Occult and the Decline of Magic', *History of Science*, 46, pp. 1-48.

Hesse, M. B. 1961. *Forces and Fields: The Concept of Action at a Distance in the History of Physics*. London: Nelson.

Kessler, E. 1988. 'The Intellective Soul', in C. B. Schmitt and Quentin Skinner (eds), *The Cambridge History of Renaissance Philosophy*, pp. 485-534. Cambridge: Cambridge University Press.

Locke, J. 1892. 'Controversy with the Bishop of Worcester', in John Locke, *Locke's philosophical works*, 2 vols. London: G. Bell.

Newton, I. 1952. *Opticks* New York: Dover.

—. 1962. 'De gravitatione et aequipondio fluidorum' [*ca.* 1668], in A. R. Hall and M. Boas Hall (eds), *Unpublished Scientific Papers of Isaac Newton*, pp. 89-169. Cambridge: Cambridge University Press.

—. 1999. *The Principia, A New Translation by I. Bernard Cohen and Anne Whitman*. Berkeley: University of California Press.

Restivo, S. 1985. *The Social Relations of Physics, Mysticism and Mathematics*. Dordrecht: Springer.

Rozemond, M. 2002. *Descartes's Dualism*. Cambridge, Mass.: Harvard University Press.

Schmaltz, T. 2008. *Descartes on Causation*. Oxford: Oxford University Press.

Schroeder, H. J. 1937. *Disciplinary Decrees of the General Councils*. St Louis, Missouri: B. Herder.

Smith, C. and Wise, M. N. 1989. *Energy and Empire: A Biographical Study of Lord Kelvin*. Cambridge: Cambridge University Press.

Stewart, B. and Tait, P. G. 1876. *The Unseen Universe, or, Physical Speculations on a Future State*. London: Macmillan. The first edition of this, in 1875, was published anonymously.

van Lunteren, F. H. 1988. 'Gravitation and Nineteenth-Century Physical Worldviews', in P. B. Scheurer and G. Debrock (eds.), *Newton's Scientific and Philosophical Legacy*, pp. 161-173. Dordrecht: Kluwer.

Westfall, R. S. 1980. *Never at Rest: A Biography of Isaac Newton*. Cambridge: Cambridge University Press.

Wilson, C. 1992. 'Euler on action-at-a-distance and Fundamental Equations in Continuum Mechanics', in P. M. Harman and A. E. Shapiro (eds), *The Investigation of Difficult Things*, pp. 399-442. Cambridge: Cambridge University Press.

Wilson, D. B. 1991. 'P. G. Tait and Edinburgh natural philosophy, 1860–1901', *Annals of Science*, 48, pp. 267–87.

Yolton, J. W. 1984. *Thinking Matter: Materialism in Eighteenth-Century Philosophy*. Oxford: Blackwell.

Chapter Six

Models and Symbols in the Understanding of Matter

Colin A. Russell,
Open University

Introduction

Ever since the simplest models were introduced, people have loved to *play* with them. This is particularly the case in chemistry, and then especially in the field of organic chemistry. Physics, too, is not exempt, Lord Kelvin famously asserting that if he could not make a model of something, he would not believe in it. Since then, of course, both chemistry and physics have reached stages when old-fashioned mechanical modelling has become quite inappropriate, if not impossible.

Mary Hesse once suggested three phases of theory evolution, which may be described as *metaphor, mechanical model,* and *mathematical representation* (Hesse 1963). How far this schema is helpful, and how far it may have theological implications, will be explored in this chapter. We may at the outset wish to add a fourth class to Hesse's scheme: *manipulative value* – for at every stage of theory evolution the issue of 'what is its use in the real world?' may well arise, and may affect our attitude to theoretical concerns.

The case of chemical atomism[1]

1. Metaphors.

Models were first seen as just metaphors: impressionistic only, imaginative and inevitably arbitrary, giving the impression that it was as though the world *had seemed* to be so constructed. In the Greek period, Democritus and others had posited a discontinuity in matter (and in time), which meant that ultimate subdivisions had to stop somewhere; and these fundamental units were thought of as 'atoms', because they were simply indivisible. No one had ever got near to them, of course, but they were a metaphorical statement as to how things might be.

From the 16th century onwards, these metaphorical units gradually assumed a stronger, more literal meaning and the metaphor was slowly transformed to a model. Robert Boyle gave such atomism a Christian spin:

> I think it probable . . . that the great and wise author of things did, when he first formed the universal and undistinguished matter into the world, put its various parts into various motions whereby they were necessarily divided into numberless portions of differing bulks, figures and situations, in respect of each other (Boyle 1996 (1686)).

Newton and others continued to hold this view, and Newton himself went further in the direction of models and referred to matter as composed of 'solid, massy, hard, impenetrable moveable particles' (Newton 1718). But there was no attempt to describe those particles, to attribute properties to them, or to use them to explain the phenomena of the newly-emerging science (apart from seeing them as participants in the new inverse square law).

Meanwhile, there were streams of 18th century thought that preferred to reckon in terms of point-centres of force to atoms. So we read in Boscovich, and later in Priestley, a non-atomistic view that would have horrified later chemists. Even Davy inclined to a tendency to regard atoms as metaphors rather than models, and he presumably passed on his scepticism to Faraday. (A good account of such 'point-centre' developments was given many years ago by L. L. White: see White 1961.)

It is tempting – but unwise – to look for simple theological correlations in these pre-Daltonian views. It is salutary to consider that Boscovich was a Jesuit priest, Priestley a Unitarian, Davy a rather uncommitted and

---

[1] The first part of this chapter is based on the Chemical Society's 1977 Grove Lecture, given by the author at the University of Swansea: 'The role of models in the evolution of chemical science'.

disenchanted Anglican, and Faraday a devout and non-conformist Sandemanian!

In the 18th century chemistry had been pursuing the 'Newtonian dream', and had certainly veered towards quantitative interpretations. Despite an enormous amount of work, little lasting progress was made for a long time. Then came Lavoisier, who discovered oxygen and who clarified the idea of chemical elements. One of his successors at the turn of the nineteenth century was a Cumbrian Quaker, John Dalton.

Dalton, founder of atomic theory

Dalton dominated Victorian chemistry with his view that an atom was still a metaphor, but came near to being a mechanical model. Its metaphorical status was guaranteed by its nature as an imaginative device, with arbitrary arrangements in space and only *ad hoc* links within molecules. This is an understanding just as described by Mary Hesse.

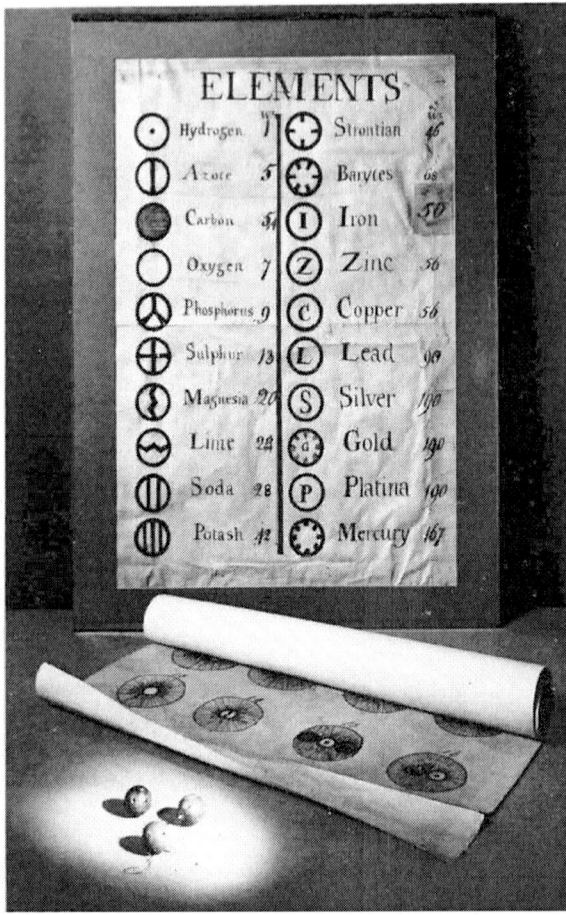

Dalton's atoms. Some of his solid spheres are in the foreground, whilst the vertical poster in the background displays his non-alphabetical symbols for different kinds of atoms.

According to Dalton, each element had its own kind of atom, with a characteristic weight and so on. Represented at first by non-alphabetical symbols (alphabetical symbols were introduced in detail by Berzelius: see Berzelius 1814), the atoms were then depicted by models, made by the Manchester engineer Peter Wart from wooden balls. Later on, atoms were facetiously described as 'square blocks of wood invented by Dr Dalton'. These models were all quite arbitrary, but imaginative. No one had seen an atom, and their presence was assumed. It has been observed that painted balls connected by rods are less like real atoms than department store mannequins are like real women.

The combinations of Dalton's atoms were developed on an arbitrary basis, which was the simplest possible. Thus water was given two atoms only, and since 8 parts by weight of oxygen combined with 1 part by weight of hydrogen, the atomic weight of oxygen taken as 8. Many other spurious atomic weights followed from this: thus 4.1 parts by weight of zinc combines with 1 part by weight of oxygen, so giving the value 4.1 x 8 = 32.8 as the atomic weight of zinc (actually half the modern value).

The next development did not come till mid-19th century, and then it was shattering in its impact.

2. Mechanical models.

Atomic models that are qualitative and detailed, specific and non-arbitrary, came to dominate the chemical scene for the last half of the 19th century. In 1865 A. W. Hofmann gave a lecture at the Royal Institution, 'On the combining power of atoms'. He used models made of coloured croquet balls on stands, remarking: 'The white balls are hydrogen, the green ones chlorine atoms; the atoms of fiery oxygen are red, those of nitrogen blue; the carbon atoms, lastly, are naturally represented by black balls' (Hofmann 1865, p. 414). This mechanical representation was of great use in teaching, but also enabled predictions to be made. Sometimes these were wrong, but this was simply because other data had been ignored. Two correct models explored by this means were ethane and ethylamine.

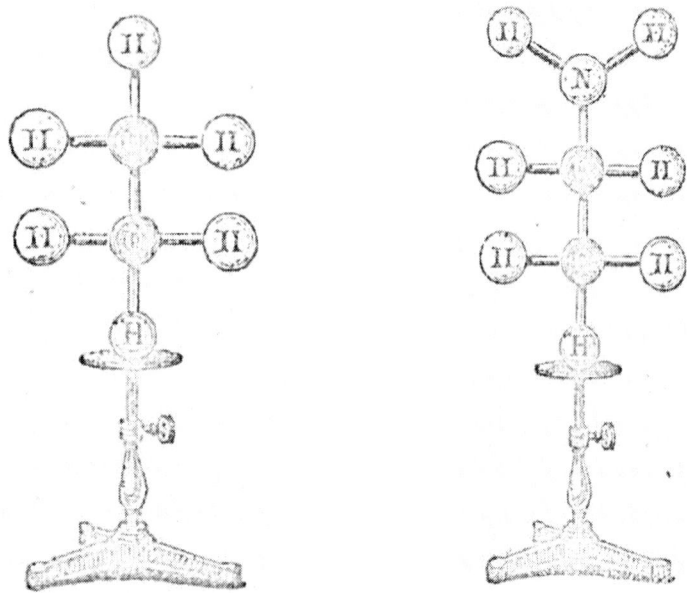

Hofmann's model atoms, made of croquet balls. In this case the models are of ethane and ethylamine.

Thus in due course the speculations of Dalton were seen as something that corresponded to reality, though with modifications. Through the work of F. A. Kekulé (1829-1896) and E. Frankland (1825-1899) a theory of chemical bonding (valency) emerged that seemed to show how and where atoms were joined together in chemical molecules. Perhaps a beginning of this theory can be located in an affirmation by Frankland which introduced the word 'bond': 'By the term *bond*, I intend merely to give a concrete expression to what has received different names from different chemists, such as an atomicity, an atomic power and an equivalence' (Frankland 1866, p. 377). Although he toyed with a fanciful analogy between a chemical bond and the force which connected members of the solar system, he soon admitted 'the bonds which actually hold the constituents together being, as regards their nature, entirely unknown' (Frankland 1870). He was still in a metaphorical mood, but this did not last long. His bonds were soon incorporated into diagrams of molecules, by means of which students could quickly obtain a very useful knowledge of organic chemistry. They filled his textbooks, and rapidly entered learned discussions of research topics.

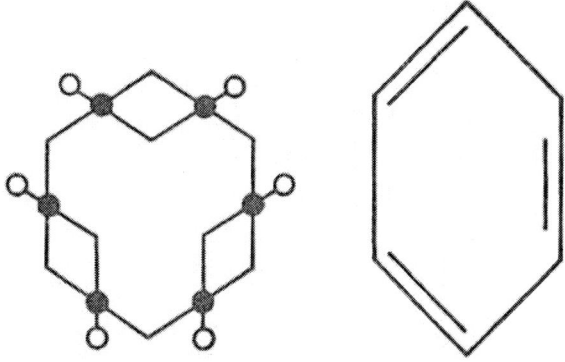

Benzene: Kekulé's first model, and modern formula

All kinds of mechanical models were soon available (through the work of Dewar, Kekulé, and others), and these models made modern formulae possible for molecules such as benzene. Eventually the Russian chemist A. M. Butlerov introduced the concept of 'chemical structure', which concretised these formulae to the extent of suggesting that each molecule had a definite structure, or arrangement of atoms and bonds, which uniquely determined its properties and which were shared by no other molecule (Mudge 1997, p. 13). The age of structural, or constitutional, formulae had arrived, and with it a clear understanding that atoms were no longer to be regarded as picturesque metaphors, but rather as real mechanical models.

However, the models developed by Hofmann suffered from a serious disadvantage. They all implied that the bonds formed by a carbon atom were in one plane, and by 1874 evidence had amassed to show this was not usually true: rather, the atom was as though situated at the centre of a regular tetrahedron, with bonds directed to its four apices. This made the –C– bond angle not 90° but rather 109° 28'. So Hofmann's model still had an element of metaphor about it, the bond angles being as arbitrary as the ideas of Dalton. The atomic structures subsequently proposed by van't Hoff immediately explained the existence of optical isomerism and much else, and were of great use for teaching.

Carbon atom in three dimensions: a tetrahedral arrangement

These were undoubtedly *mechanical* models. They were also used to construct semi-scale models of actual molecules and this led to phenomena like steric hindrance being predicted and interpreted. Above all, their application to cyclohexane rings led Sachse (1890, p. 1363) to the prediction of two forms of this (boat and chair). Mohr (1918, p. 318: 1922, p. 316) later showed how this meant that decalin must also exist in two forms, and that these would be separable; and indeed they *were* separated, by Hückel, in 1925. Thus the foundation of conformational analysis was laid, together with many other aspects of organic chemistry concerned with the actual three-dimensional shapes of molecules.

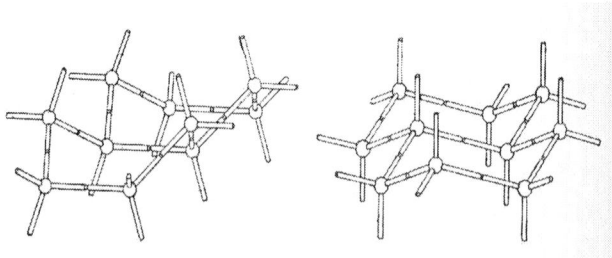

Decalin in 'boat' and 'chair' conformations

3. Mathematical statements.

Organic chemists might have continued using models like these for years, as indeed they still do in many fields. But a further phase was being recognised. The discoveries of both the electron and of the nuclear atom soon led to quantum mechanics and the Schrödinger wave equation (see the chapter by Ruth Gregory in the present volume). For some chemists, and for most physicists, a new era had begun: the atom could be expressed mathematically, through the solution of complex differential equations.

Was it a particle or a wave? Bohr imported the idea of complementarity from the ideas of the theologian Kierkegaard. Since then, the concept has been applied to many classical dilemmas, not always correctly. From our new understanding of matter, complementarity has become a mode of thought that also enables the Arminian and the Calvinist to get on well together: not *either/or* but *both/and*.

Our confident understanding of chemical atomism has been severely dented, and the depths of our ignorance about matter further exposed. As Coulson once said:

> I described a bond, a normal simple chemical bond; and I gave many details of its character (and could have given many more). Sometimes it seems to me that a bond between two atoms has become so real, so tangible, so friendly that I can almost see it. And then I awake with a little shock; for a chemical bond is not a real thing; it does not exist; no one has ever seen it, no one ever can. It is a figment of our own imagination (Coulson 1955, p. 2084).

The whole of modern chemistry rests upon a theory of atomism. But it is expressed most fully by what amounts to a series of equations, a truly mathematical representation.

In fact, all three of the phases described by Hesse overlap, and it is sometimes hard to draw the line between them. In practice all are followed by another concern (which in one form or another has been present all the time): manipulative value.

4. Manipulative value.

Hesse's threefold model may even be applied to the theory of evolution. But there is another criterion in theory development which is of great importance. Scientific theories will only be accepted in a long run if they are additionally seen as having actual or potential *manipulative value*, and that is our fourth phase, to be added to Hesse's three. The case of atoms can be dealt with quite briefly.

First, in chemistry itself a fruitful concept of matter is now inconceivable without the idea of atoms (despite the exertions of Nernst and a few others). In fact they are largely taken for granted, and that is the highest accolade of acceptance. The whole of chemical structure and reactivity depends on a clear perception of atoms and bonds. The magisterial theories of Ingold and others in the mid-20th century lays organic reactions open to a gaze that is fully committed to atoms, though the mechanisms of a number of reactions (like molecular rearrangements) depend on a different kind of atomism from that disseminated by Frankland and his colleagues. Who of them would have imagined that one carbon atom could be bound simultaneously, though only briefly, to five ligands at once? But it is because of their explanatory power that atoms, sometimes in new forms, seem to be the essence of chemical theory. And of course their use in chemistry is far from merely explanatory. The number of useful compounds which may be created through their combination can be numbered in millions, and their evident utility as a means of understanding such compounds is obvious, from the creations of the pharmaceutical laboratory to the physiological and criminological applications of DNA. Indeed, scarcely any biological phenomenon is conceivable today without an atomic or molecular explanation at its heart.

The manipulative functionality of atoms may be also displayed in the immense applicability of nuclear transformations. Since we now know that atoms need not be eternal, we can manipulate their very selves through the phenomena of radiochemistry. These include the production of radio-isotopes for use in medicine, as biological tracers and in certain specific therapies. On a larger scale, the fission or the fusion of nuclei of these same atoms can yield power undreamt of by previous generations. In a more sinister way, these processes may be also used to create atomic weaponry. None of the advances, beneficent or not, would be possible without a clear (if ever-changing) acceptance of the theory of atoms.

Now we turn from atomic theory to a very different set of ideas, those relating to climate change. We shall find that precisely the same four steps may be discovered in the evolution of theories in this area, too.

A Case Study: carbon dioxide and global warming

1. Metaphor.

It was a French mathematician, J.-B. Fourier, who first pointed out that the atmosphere acted like a transparent glass cover of a box. He called this the 'hothouse effect', but it was still a splendid metaphor. This idea was taken up in 1896 by Arrhenius, and later a few others. But they were largely ignored. 'Greenhouses' were apparently first referred to by John Evelyn in 1664. With a rising popularity of growing plants under glass, the climatic effect became known as the 'greenhouse effect'; but the first specific reference to a greenhouse effect that I have found is by G. T. Trewartha in a book of 1937 entitled *Introduction to weather and climate*. In this book Trewartha refers to 'the so-called greenhouse effect of the atmosphere'. Occasional references occur in the 1960s. Then, in November 1968, an *Observer* article proclaimed that 'carbon dioxide . . . is responsible for a "greenhouse" effect which allows in heat from the sun, but prevents it escaping back into space'. By the 1980s the metaphor was in common usage, although it was still just an engaging, if worrying, metaphor (see Intergovernmental Panel on Climate Change 2001a).

2. Mechanical models.

For hundreds, probably millions, of years our Earth has been relentlessly bombarded with everything the sun can throw at it. Solar rays include heat, together with all kinds of waves and particles we would rather not have. We also have an atmosphere that can let most of the sun's rays through, though the ozone layer filters off the harmful 'hard' ultra-violet rays. One consequence of our receiving all that heat from the sun is that we manage to keep (mostly!) comfortably warm, though the Earth also gets some heat from naturally radioactive minerals in its interior. We now know that about 30% of the sun's radiation is reflected back at it, mainly by clouds, but there is one snag. The carbon dioxide in our air helps to absorb natural radiation *from* the Earth, so we get warmer than would otherwise be the case. (In its absence the expected average temperature would be around -18°C, but in fact the average is about +15°C.) In other words, it was suggested that CO_2 acts just like the glass in a greenhouse, and hence carbon dioxide is said to exert a 'greenhouse effect'. The involvement of carbon dioxide was assumed, from studies of radiation effects, cosmic chemistry and related sciences.

Other greenhouse gases exist, such as methane (from land-fills), water vapour and ozone. Much study has taken place of free radical reactions in the stratosphere. A mechanism for this effect was proposed, though this is too complex for discussion here. The 'greenhouse effect' now has the status of a model, with correlations between what is observed, and what the model would predict. It is no longer a hand-waving metaphor, of great charm no doubt, but with no 'mechanical' basis.

Now, CO_2 is the greenhouse gas that has been most studied, and there is no doubt that in the past higher amounts has meant higher temperatures.

Does this mean evil portents for the future? The answer to this question emerged at an international conference.

3. Mathematical expressions.

A series of mathematical models was worked out and studied at an Intergovernmental Panel for Climate Change, with several hundred leading scientists from all over the world. An astonishing consensus emerged about the past and future rise in temperature, and with it many different equations. The following graphs were published, showing past warm-up and the predictions for the future:

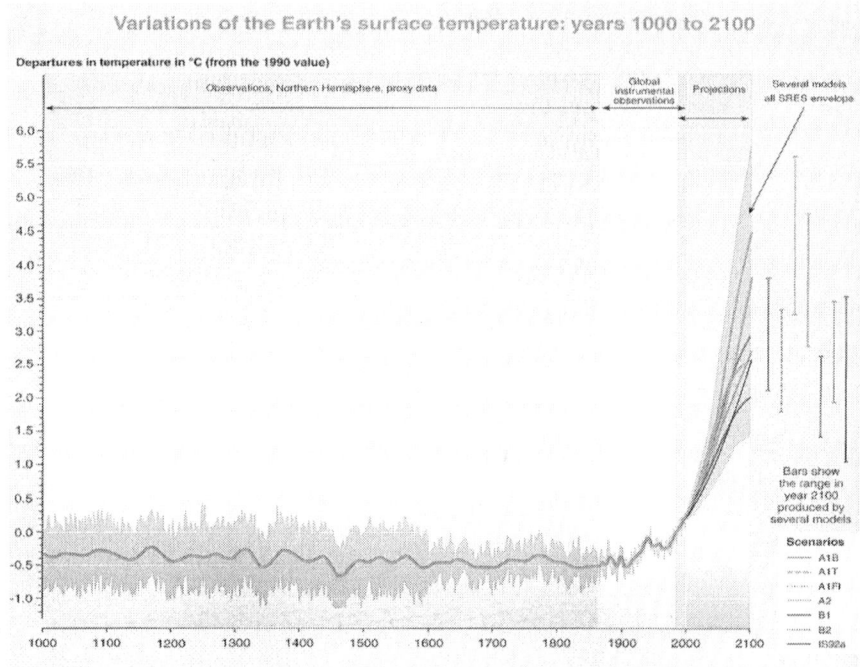

Variation with date of CO_2 atmospheric concentrations

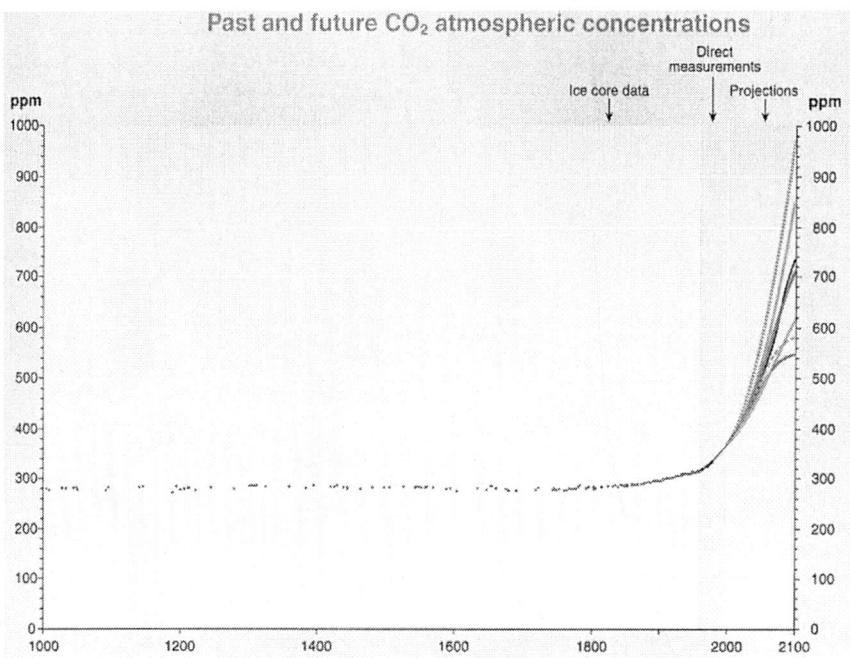

Variation with date of earth's temperature

(Graphs from Intergovernmental Report on Climate Change, 2001a, pp. 33
and 34. Reproduced by permission of Sir John Houghton.)

The fuzziness of the projections is due to the application of different
equations, but the trend is inescapable. The temperature rises when the
tropospheric concentration of carbon dioxide also rises. The 'greenhouse
effect' now has a solid mathematical basis.

This case not only illustrates the development of a scientific theory,
where mathematicisation is likely to predict the way forward. It was a
detailed study of equations and graphs that led the International Panel for
Climate Change (IPCC) to its virtually unanimous agreement about the
cause of the greenhouse effect. All their work was governed by a concern
to do something about it. In other words, this greenhouse doctrine had
important applications. And these would not have been evident without
recourse to extensive mathematics.

4. Manipulative value.

Clearly carbon dioxide is at the heart of the problem. Its role in our atmosphere is sufficient grounds for interest and concern. In the last 20 years or so, the ontological status of the so-called 'greenhouse effect' has changed remarkably. Matter is now seen as needing manipulation if we are to survive. This is, of course, a subject for politicians, journalists, and pressure groups, such as those which invite us to be 'Friends of the Earth'. Theologically, it is striking that Biblical data do not speak at all of matter being semi-divine, or even alive (they actually imply the opposite). So the Earth needs 'Friends' no longer in any mystical sense, but as those who will manipulate it in a beneficial manner, mainly by avoiding waste and by keeping CO_2 emissions as low as possible.

There is a remarkable theological implication here, and in the last 20 years or so many in the Christian church have found themselves in surprising alliances. Some years ago William Temple said of Christianity that it was the most materialistic of all religions (Temple 1960, p. 478). Matter must be seen to matter. Only if this world of atoms and molecules is not ours, but God's, can any sure future be envisaged. To care for it is our mandate as human beings. Christianity maintains that this is not just a question of human survival. Only if this world of atoms and molecules is not ours, but God's, can any sure future be envisaged. The Old and New Testaments portray the earth as not ours but the Lord's; moreover, to care for it is our mandate as human beings, from Genesis 2:23 onwards. We are instructed to care for our planet as human stewards. Not only will this determine the Master's 'well done' or otherwise, but our response will also affect the well-being of our descendants, and of the poor and underprivileged already in this world. It is noteworthy that of the multitude of books which has been written on climate change, some come from a directly Christian perspective (see, for example, J. Moltmann 1985, Spencer and White 2007, Prance 1996, Houghton 2007, Berry 2007; and, for a more general account, Russell 2008).

Christians have found that in our lifetime matter has become not profane, but sacred, because it is the creation of the cosmic Christ. This may be seen through Christ's being at work in the storm (when even wind and waves obeyed Him), and also in His care for the sick. Here matter is given huge religious meaning: as Peacocke put it, 'The world of matter is seen both as expressing and revealing the mind of God as Creator, and of expressing His purposes' (Peacocke 1979, p. 290). To see this most forcibly we need perhaps a fresh understanding if the Easter narrative as well. In this we can perhaps see that the world of matter will be transformed into a form related to, but different from, the world of matter we experience today. As Tom Wright has written, this gives us tremendous hope for the future (Wright 2007). It has also given us an unprecedented challenge for today, to transform our metaphors, mechanical models, even their mathematical expressions, into our involvement in the divine plan for the cosmos.

Bibliography

Berry, J. (ed.) 2007. *When enough is enough*. London: IVP.

Berzelius 1814. *Ann. Phil.*, iii, 51.

Boyle, R. 1996. *A free enquiry into the vulgarly received notion of nature*. Ed. E. B. Davis and M. Hunter. Cambridge: Cambridge University Press (originally published 1686).

Coulson, C. A. 1955. 'The contribution of wave mechanics to chemistry', *J. Chem. Soc.* pp. 2069-2084.

Frankland, E. 1866. *J. Chem. Soc.*, vol. 19, 377.

—. 1870. *Lecture notes for chemical students, vol. I*. London.

Hesse, M. B. 1963. *Models and Analogies in Science*. London.

Hofmann, A. W. 1865. *Proc. Royal Inst.*, vol. 4.

Houghton, J. 2007. *The search for God: can science help?* Second edition. Oxford: Lion.

Intergovernmental Panel on Climate Change, 2001a. *Synthesis report*. Cambridge: Cambridge University Press.

—. 2001b. *The scientific basis*. Cambridge: Cambridge University Press.

Mohr, E. 1918. *J. Prakt. Chem.*, vol. 98.

—. 1922. *J. Prakt. Chem.*, vol. 103.

Moltmann, J. 1985. *God in creation: an ecological doctrine of creation*. London: SCM Press.

Mudge, F. B. 1997. 'The development of greenhouse theory of global climate change from Victorian times', *Weather*, vol. 52.

Newton, I. 1718. *Opticks*. London.

Peacocke, A. R. 1979. *Creation and the World of Science*. Oxford: Clarendon Press.

Prance, G. 1996. *Earth under threat*. Wild Goose.

Russell, C. A. 2008. *Saving Planet Earth: a Christian response*. Milton Keynes: Authentic Media.

Sachse, H. 1890. *Berichte*, vol. 23.

Spencer, N. and White, R. 2007. *Christianity, climate change and sustainable living*. London: SPCK.

Temple, W. 1960. *Nature, Man and God*. London: Macmillan.

Trewartha, G. T. 1937. *Introduction to Weather and Climate*. London: McGraw Hill.

White, L. L. 1961. *Essay on atomism from Democritus to 1960*. Nelson.

Wright, N. T. 2007. *Surprised by hope*. London: SPCK.

CHAPTER SEVEN

A RESPONSE TO COLIN RUSSELL'S
MODELS AND SYMBOLS
IN THE UNDERSTANDING OF MATTER

MICHAEL POOLE,
KING'S COLLEGE, UNIVERSITY OF LONDON

Colin Russell's paper, *Models and Symbols in the Understanding of Matter,* raises a number of issues. It rightly requires some explication of what a 'model' in this context means, how it relates to Mary Hesse's first phase of theory development – that of metaphors – and why models are so widely employed in science and elsewhere.

In Russell's paper, four phrases caught my eye:-

1. **'Robert Boyle gave atomism a Christian spin'** (p. 68). This was the first hint in the paper of a religious dimension. The question, 'Why should there be a religious dimension?' can be answered by responding that there is often a quest for something more fundamental underlying physical realities. Robert Boyle's Christian beliefs underpinned his quest. Boyle objected to the Greek notion of the World as semi-divine because he considered it idolatrous and because the 'veneration wherewith men are imbued for what they call nature' was 'a discouraging impediment' to experimental philosophy (cited in Poole 1995, p. 50).

Perhaps a more forceful example of this interaction between model-making and religious belief would be Boyle's mechanistic model for the World. Non-believers in God could, and perhaps did, make similar suggestions, but once the mechanism was identified as a clock, with reference to the huge timepiece in Strasbourg cathedral, the matter of agency and of a 'clockmaker' took a key place. The fact that the metaphor later backfired offers a cautionary tale about using metaphors.

2. 'Bohr imported the idea of complementarity from the ideas of the theologian Kierkegaard. Since then, the concept has been applied to many classical dilemmas, not always correctly. From our new understanding of matter complementarity has become a mode of thought that also enables the Arminian and the Calvinist to get on well together: not *either/or* but *both/and*' (p. 75). I found some difficulty with aspects of this statement, particularly the clause 'complementarity has become a mode of thought that also enables the Arminian and the Calvinist to get on well together.'

Complementarity is a thorny topic. Karl Heim (1953), C. A. Coulson (1955), Donald MacKay (1974) and W. H. Austin (1976) were among those who have grappled with it. The matter at issue is whether the Principle of Complementarity, enunciated by Bohr, has any application outside of microphysical phenomena. Bohr himself thought so, and many apologists since that time have claimed 'complementarity', or a modified form of it, as a means of bringing together apparently disparate subjects like science and religion. One approach, exemplified by C. A. Coulson in his *Science and Christian Belief* (1955), consists of an attempt to justify speaking of science and religion as 'complementary' as an extension of the use of the idea of complementarity in physics. Another kind of approach, developed by Donald MacKay (MacKay 1974), attempts to define a *logical relationship* called 'complementarity' which does not depend on Bohr's initial use of the concept, although the wave-particle duality is considered to exemplify it.

Of all variants of the first approach, Barbour points out, '… the use of the Complementarity Principle outside physics is *analogical not inferential*. There must be independent grounds for justifying in the new context the value of two alternative sets of constructs' (Barbour 1966, pp. 292ff). Furthermore, in physics the Complementarity Principle 'refers to *different ways of analyzing a single entity* (such as an electron)' (Barbour 1966, p. 293); and although it *might* be possible to extend the principle to 'mind-and-brain', it would not be legitimate to apply it to science and religion, since 'science and religion are not simply two views of one world – unless one subscribes to pantheism' (Barbour 1966, p. 293).

Another dissimilarity is that in Bohr's original usage the wave-particle descriptions occurred within the same universe of discourse. Science and religion concern different universes of discourse. Even these few points are fatal to attempts to establish some kind of logical relationship between scientific and religious accounts of the world on the strength of Bohr's Complementarity Principle. If scientific and religious accounts are to be classed as 'complementary', then a different approach to the meaning of the word 'complementary' is needed. Such an approach was attempted by MacKay, but is beyond the scope of this response.

3. **'Manipulative value'** (p. 67 *et seq.*). Manipulation sounds sinister, and the use of metaphor can bring strong pressures with it. James Lovelock, author of the Gaia hypothesis, wishes to persuade us that the Earth is alive and attempts to put across this counter-intuitive idea with the carefully chosen metaphor of a giant redwood tree, saying:

> You may find it hard to swallow the notion that anything as large and apparently inanimate as the Earth is alive. Surely, you may say, the Earth is almost wholly rock and nearly all incandescent with heat ... the difficulty can be lessened if you let the image of a giant redwood tree enter your mind. The tree undoubtedly is alive, yet 99 percent is dead. The great tree is an ancient spire of dead wood, made of lignin and cellulose by the ancestors of the thin layer of living cells that go to constitute its bark. How like the Earth, and more so when we realize that many of the atoms of the rocks far down into the magma were once part of the ancestral life from which we all have come (Lovelock 1991, p. 27).

However, I don't think that Russell is using manipulation in this rather forceful way, but rather to present a rational case for the development of a theory. But even if manipulation can be benign, is there another danger – that of being misled? What about the vain search for an *aether* for the propagation of light, because both water waves and sound waves require a material medium for their propagation? What about the way in which the evolutionary model of biology has become a 'control model' for countless other phenomena, from motor cars to postage stamps? Then again, how about the weaknesses in the solar system model of the atom? The philosopher R. B. Braithwaite comments:

> Thinking of scientific theories by means of models is always as-if thinking; hydrogen atoms behave (in certain respects) as if they were solar systems each with an electronic planet revolving round a protonic sun. But hydrogen atoms are not solar systems; it is only useful to think of them as if they were such systems if one remembers all the time that they are not. The price of the employment of models is eternal vigilance (Braithewaite 1968, p. 93).

4. **'Christians have found that in our lifetime matter has become not profane, but sacred, because it is the creation of the cosmic Christ. ... Here matter is given huge religious meaning'** (p. 80). How does matter *become* sacred? Can it be sacred anyway? A dictionary definition

of 'sacred' as 'dedicated to religious use; dedicated or dear to a divinity; set apart' suggests it can, in some senses of the word. But what about 'profane' as 'irreverent; secular; not sacred'?

Then again, is it possible for matter to 'speak to us' in such a way as this? As we might expect, there have been a variety of responses to this question.

(i) Tennyson seemed to think it could. In his 'In Memoriam', he wrote of humankind,

'Who trusted God was love indeed
And love Creation's final law –
Tho' Nature, red in tooth and claw
With ravine, shriek'd against his creed' (Tennyson 1973, pp. 35-6).

(ii) The Apostle Paul in his letter to the Romans (1:20) says that: '... since the creation of the world God's invisible qualities – his eternal power and divine nature – have been clearly seen, being understood from what has been made'. However, Paul also notes the suffering found in the world. In the same letter (8:22) he observes: 'We know that the whole creation has been groaning as in the pains of childbirth right up to the present time.' The problem of *theodicy*, hinted at here, of reconciling belief in a loving God with a travailing world is a thorny one, which has recently been explored by Southgate (Southgate 2008).

Bibliography

Austin, W. H. 1976. *The Relevance of Natural Science to Theology.* London: Macmillan.
Barbour, I. G. 1966. *Issues in Science and Religion.* London: SCM Press.
Braithwaite, R. B. 1968. *Scientific Explanation.* Cambridge: Cambridge University Press.
Coulson, C . A. 1955. *Science and Cristian Belief.* London: Fontana.
Heim, K. 1953. *Christian Faith and Natural Science.* London: SCM Press.
Lovelock, J. 1991. *The Ages of Gaia: A Biography of Our Living Earth.* Oxford: Oxford University Press.
MacKay, D. M. 1974. ' "Complementarity" in Scientific and Theological Thinking'. *Zygon* 9 (3), pp 225-244.
Poole, M. W. 1995. *Beliefs and Values in Science Education.* Buckingham: Open University Press.
Southgate, C. 2008. *The Groaning of Creation: God, evolution and the problem of evil.* Westminster John Knox Press.
Tennyson, A. 1973. *In Memoriam.* New York: Norton.

CHAPTER EIGHT

MATTER:
AN ISLAMIC PERSPECTIVE

M. B. ALTAIE,
YARMOUK UNIVERSITY, JORDAN

In this essay I wish to comment on philosophical understandings of matter chiefly from an Islamic perspective, elaborating on the question of whether it is sacred or profane. For this purpose I will specifically present the viewpoint of Kalām, which might be taken to represent the view of large group of Muslim intellectuals. This response will help complement the historical aspects of this volume, as well as enrich the philosophical side of the subject.

With the establishment of the Abbasid caliphate, translations of texts on various sciences from Greek, Syriac, Persian and Sanskrit sources became available in Arabic. This resulted in the creation of new schools of thought in the Islamic traditions: in addition to the earlier schools of grammarians and poets, traditionalists, commentators, historians, and Sufi ascetics, all of whom relied almost entirely upon the Islamic revelation for their knowledge, there now began to appear new schools that also drew from non-Islamic sources (Nasr 1978, p. 12).

During the third century the Islamic spirit began to be crystallized into permanent forms, as reflected in the formation of the schools of Kalām, law and Sufi brotherhoods. The flourishing of the various arts and sciences, as well as of philosophy, reached its climax in the fourth and fifth Islamic centuries. During that time, Muslims had the historic opportunity to re-visit the same problems which had been tackled by preceding thinkers, mainly Greek and Hindu. In this endeavor the Muslim theologians, Sufi, philosophers and Mutakallimūn expressed their views about cosmological questions and established their cosmological doctrines. (The Mutakallimūn are a group of Muslim thinkers and theologians who first appeared during the 8th century A.D. and who believed that the basic principles of faith should be realised through reason. They devoted themselves to structuring Islamic religious beliefs in accordance with a rational approach. Their works laid the foundations for the Islamic philosophy of nature. Many of the subjects they studied are similar to those dealt with in today's science and religion debates, and many of their arguments resemble modern ones as far as metaphysical questions are concerned.)

It could be said, without much reservation, that Muslim philosophers like Al-Farabi, Ibn Sina (Avicenna) and Ibn Rushd (Averroes) drew their views mainly from original Greek sources, trying to construe the philosophical doctrines within an Islamic perspective. Much to their disappointment, the theologians and Mutakallimūn were to stand on the opposite side, trying to refute the philosophical approach, whilst the Sufi had their own ascetic, non-rational views of God, matter and mankind. It can also be said without much reservation, and in full agreement with Richard Walzer, that 'the Mutakallimūn followed a methodology that is distinct from that of the philosophers in that they take the truth of Islam as their starting point' (Walzer 1970, p. 648). William Craig has elaborated this point further, by saying that

> The main difference between a *Mutakallim* and a *Failasuf* lies in the methodological approach to the object of their study; while the practitioner of Kalām takes the truth of Islam as his starting-point, the man of philosophy, though he may take pleasure in the rediscovery of Qur'anic doctrines, does not make them his starting-point, but follows a method of research independent of dogma, without, however, rejecting the dogma or ignoring it in its sources (Craig 1979, p.17).

When dealing with the history of atomism, many Western authors jump from Democritus in the 5th century B.C. to John Dalton in the 19th century A.D., ignoring the history in between. It is known that many Muslim thinkers dealt with philosophical questions about space, time and matter within a similar context as that dealt with by the Greeks. Indeed, the meaning of 'matter' and the notion of atomism were basic subjects of Kalām. The Mutakallimūn discussed thoroughly the question of whether the world is continuous or discrete, and in one important and original contribution to the notion of atomism they devised some abstract concepts on this topic. In fact, their contributions were so original that they have pushed Wolfson and other authors to question the sources of the abstract concepts and properties which were associated with the Islamic version of atomism (Wolfson 1976). However, after his thorough investigations of the prime sources of Islamic atomism, Shlomo Pines clearly asserted that any attribution of these sources to Hindu or Greek origins is doubtful (Pines 1946). In fact, the atomic theory of Kalām was an integral part of a whole philosophical theory of the world, in which the temporality of the world, the discreteness of space, time and matter, the continued re-creation of accidents, the indeterminism of the world, and the integrity of spacetime, played complementary roles to explain the existence and phenomena of the world, in conformity with the necessity of a persistent Creator (Altaie 2007). This leads to an holistic concept of the world, by which every part of it is, in one way or another, entangled with every other part.

The main principles of Daqiq al-Kalām

Despite the differing views expressed by the Mutakallimūn who belonged to different schools of Kalām, we find that most of these views do follow some common basic principles that were adopted to understand nature. I have identified the following five principles (Altaie 1994).

1. The creation of the world.

The best available account of this doctrine was given by Al-Ghazali in his celebrated book *Tahafut al-Falasifa* (Al-Ghazali 2000). According to the Mutakallimūn the world is not eternal, but was created at some finite time in the past. Space and time have no meaning, and did not exist before the creation of the world. Despite the fact that some of the Mutakallimūn believed that the creation of the world took place out of a pre-existing form of matter, the dominant view of the Mutakallimūn in this respect was that creation took place *ex nihilo*, i.e. out of nothing (Al-Alousi 1980, p. 59: Wolfson 1976, pp. 359-372). Accordingly, they considered every constituent of the world to be temporal. William Craig (1979) re-framed this doctrine in a more modern context, and used it as a proof for the existence of God.

2. Discreteness of natural structures.

The Mutakallimūn believed that all entities in the world are composed of a finite number of fundamental units. Each was called *Jawhar* (the substance): it is non-divisible and has no parts. The *Jawhar* was thought to be an abstract entity that acquires its physical properties and value when occupied by a character called *'Aradh* (i.e. the accident). These accidents are ever-changing characters. Discreteness applies not only to material bodies but also to space, time, motion, energy (heat), and all other properties of matter (Altaie 2005). Some authors have tried in vain to relate the Islamic concept of the atom to that of Greek or Hindu thinkers, although Wolfson (1976, pp. 471-472) pointed out basic differences between the Islamic atom and the Greek one. Since the Islamic atom has no magnitude, and because the number of atoms in the world is finite, it is unlikely that the Muslims have taken this idea from any other context: the Islamic atom has genuinely different properties which make it distinguishable from others (Pines 1946).

3. Continual re-creation and an ever-changing world.

Because God is the active creator of the world, and because God acts continuously to sustain the universe, therefore the world has to be re-created at every moment (Al-Juwayni 1969, p. 159). This re-creation occurs with the accidents, not with the substance; but since the substances cannot be realized without being attached to some accidents, therefore the re-creation of the accidents effectively dominates the substances too. By such a process, God acts as the sustainer of the world.

This principle can be utilized to explain the indeterminism of the world. Indeed, in a separate chapter in this volume (chapter 3) the application of the notion of re-creation is explored.

4. Indeterminism of the world.

Since God possesses absolute free will, and since He is the personal creator and sustainer of the world, then He is at liberty to take any action He wishes regarding the state of the world or its control. Consequently, the laws of nature that we observe have to be probabilistic not deterministic, and physical values have to be contingent and undetermined. This is how the Mutakallimūn deduced the indeterminacy of the world; and this resulted in their rejection of the existence of natural causality, because nature, according to the Mutakallimūn, cannot have any sort of will. The Mutakallimūn also rejected the four basic Greek elements (Al-Baqillani 1987), and denied the existence of any kind of self-acting property belonging to those elements. This is a central argument in Kalām for the proof of the need for God; if nature is blind, then no productive development would be expected.

5. Integrity of space and time.

The Mutakallimūn had the understanding that space has no meaning on its own. Without having a body, we cannot realize the existence of space. The same applies to time: this cannot be realized without the existence of motion, which needs a body to be affected (Altaie 2005). This connection of space and time is deeply rooted in Arabic. Therefore, according to Kalām, neither absolute space nor absolute time exists (Ibn Hazm 1983, p.

75). This generated their understanding of motion as discrete, and the trajectory of motion as being composed of neighboring '*rest-points*' (Al-Ash'ari 1980, p. 21-25). Accordingly, they maintain that a body is perceived to be moving faster than another only because the number of rest-points along its trajectory is small compared to those along the trajectory of the other.

However, the Mu'tazilite al-Nazzam believed that motion on the microscopic level takes place in discrete jumps called '*tafra*'. Max Jammer considered this understanding of al-Nazzam to be the oldest realization of a quantum motion: he wrote, 'In fact al-Nazzam's notion of leap, his designation of an analyzable inter-phenomenon, may be regarded as an early forerunner of Bohr's conception of quantum jumps' (Jammer 1974, p. 259).

The Current Status of Quantum Theories of Matter

We need to understand the true implications of the 20th century sciences as much as we need to understand the original doctrines of religion. Concepts proposed by quantum theorists and the mathematical structure of quantum mechanics are still in need of deeper understanding. The meaning of an 'operator' in quantum mechanics is as obscure as the meaning of the unpredictability of measurements. The role played by the mathematical entities called 'imaginary quantities' in physics, although they cannot be measured directly, is also something worth studying on the conceptual level, in order to understand more of its practical naturalistic meaning. (For example, Hawking and Hartle, working on the wave function of the universe, declared in 1982 that the universe could have been in a state of an infinitely-extended imaginary time before the big bang; but imaginary time is not a physical time that can be measured.)

In theoretical physics, most of us play the game of generating equations that sometimes do not have clear explanations. An example of this is string theory. In general relativity, and in curved spacetime physics, we are not yet able to understand the full meaning and implications of space-like universes, for example. For this reason, many of the black hole physicists were taken by surprise by Stephen Hawking's declaration, at the 2008 GR17 conference held in Ireland, that information is not completely lost when a particle falls into a black hole. In cosmology, despite the pre-eminence of the big bang theory, we are still far from deciding whether the universe had a start in time, or whether it had an infinite extension in the past. The point singularity containing all the matter and energy that exists in our universe stands as not only an epistemological challenge, but as an ontological dilemma too. To be sure, science is firm and strong on the practical side of the story, but it is still far from reaching a conclusion on the theoretical side. That is why we should perhaps not speculate too much. Instead, we should have some fixed basic principles and doctrines, some sort of an epistemic paradigm, while finding our way through science and religion. Generally, it might be said that theoretical physics now is in a situation similar to that of the classical physics at the end of the 19th century. A breakthrough is needed to reform all modern physics into one integrated body of knowledge, by which relativity theory and quantum mechanics merge into one coherent account that will explain space, time and matter in clearer ways.

Some issues of contemporary importance

In this section I will discuss some currently 'hot' issues in the dialogue of science and religion, adopting the arguments of Kalām as the background of my suggestions. My aim is to put forward some ideas in dealing with these issues, and to explore what the Islamic Kalām may have to say about them.

1. The laws of physics and the laws of nature.

Most of us refer to the laws that are discovered in physics as the laws of nature. In fact this is a subtle point, and would imply some sort of a belief that we may or may not accept. To admit the existence of 'natural laws' may implicitly be to say that nature presents itself according to reliable sets of rules, which control its behavior. However, this may also mean that the laws of nature are some intrinsic properties within nature that make it behave spontaneously independent of any thing beyond it. Ancient philosophers assumed such intrinsic properties, and today's atheist scientists assume the same. This led in the past to a kind of reductionism: this can be observed even in Aristotle's doctrines, which assume that God is just a 'prime mover'. In due course, this led to the 'God-of-the gaps' approach, and to the concession of a role for God only if a beginning in time for the universe is realized. Therefore, as soon as Hawking deduced a way to avoid the temporal singularity in the history of the universe, he immediately went on to question the role of the creator. In fact, Hawking's question would be inevitable for any one who sees no role for God except as a prime mover: if there is no beginning, this will eliminate the role of the prime mover.

Some physicists, like Steven Weinberg, are skeptical regarding any divine interaction with the world, and wish to see God always acting miraculously. During a debate with John Polkinghorne, Weinberg said: 'At any moment we may get evidence of a supernatural supervisor of the universe. I mean suddenly in this auditorium a flaming sword may come and strike me for my impiety; and then we will know the answer' (c.f. Polkinghorne and Welker 2001, pp. 12-13). However, a miraculous universe is more likely to be chaotic, and a chaotic universe will be in less need for God (although such a need cannot be fully eliminated). If the universe runs miraculously, the task of assuming the absence of an organizing and controlling global force would be easier. On this point, it seems to me that the argument of Weinberg is self-defeating: a fiery sword will suddenly appear in the auditorium to hit Steven Weinberg if and only if the world is working completely randomly.

2. Glimpses of divine action.

Our understanding of divine action in the world will shape our understanding of the divine attributes and capabilities, and consequently will shape our understanding of God. Therefore, it is of the utmost importance to be cautious in considering our limited intellectual capabilities and our corrigible scientific knowledge.

Quantum theory provides us with a new realization of the world, through new concepts and principles that seem to transform our conception of nature and to make it more abstract. The observation that particles have wave properties has weakened concepts of locality, and produced the result that the physical measurements of some parameters are undetermined. The Heisenberg uncertainty principle, and the notion of virtual particles that was derived from it, allow for the postulation of invisible and directly non-measurable virtual worlds surrounding us. The notion of a vacuum is, accordingly, different from the customary notion of nothingness, since virtual particles, which can be transformed into real particles, are normally thought to exist.

The so-called 'causal joint' in an explanation of divine action is now often sought in the quantum description of the world. At this point we should remember that the quantum description of the world, even at the macroscopic level, is physically more accurate than the classical description. So, indeterminism and probabilistic measurements underlie the reality of our physical world. On the other hand, quantum descriptions of the physical world demand the presence of 'operators' that would affect the action of measurement, or any movement of the system. Such operators are understood to be mathematical entities within the structure of quantum theory, and physical observables are the corresponding expectation values of those operators.

The full description of the world on the smallest scale will require the quantization of spacetime, a step that would reformulate the whole structure of both quantum theory and general relativity. Some basic concepts may have to be altered accordingly, and we should therefore be careful about drawing final conclusions regarding our models of God, since God should be independent of all this. That is to say, any comprehension of divinity and of divine action should be independent of the details of scientific theories: we should take the evidence from science only in as much as it would guide us in a rational understanding of divinity. But it should always be remembered that comprehending divinity is more a matter of faith than a mathematical exercise. No-one can prove or disprove the existence of God mathematically, and since God is not a physical entity, no one could prove or disprove the existence of God through physical discoveries.

The Islamic description of God in the Qur'an is more abstract than the description given in the Bible. However, we should admit that both the Qur'an and the Bible give some descriptions of the divine attributes that may look, at first sight, to be self-contradictory. This stems from the fact that holy books are not scientific books, but make frequent use of metaphor.

3. Intelligibility of the physical world.

Perhaps this is the most important issue in reconciling science and religion. The universe seems to follow logical patterns of causality and lawfulness. This leads to the belief that nature is unfolding itself, with no need for an external driver. However, this may lead also to the belief that the universe has a sort of cosmic mind that is driving it from within. This was the type of God that Spinoza conceived, and in which Einstein believed. However, it remains problematic to see how such cosmic order could generate a global cosmic consciousness solely from within the cosmos itself. It remains a challenge to understand why this intelligible world is organized in such a way as to behave lawfully. Therefore, it seems that the lawfulness of the world is a strong indication of its having both a purpose and some final destiny.

One might suggest that objects in the world do have some intrinsic nature; however, there can be no clear reason for this nature being a stand-alone property, taking into consideration the fluctuating character of all physical quantities demonstrated by quantum physics. Here the principle of continuous re-creation found in Kalām comes into action. Fluctuations are caused by continual re-creation, justifying at the same time the divine intervention driving the world through its re-created properties.

Such divine intervention need not be miraculous, since such miraculous intervention would render the world unintelligible, as already mentioned. The Qur'an has stressed this fact in several verses. For example we read, '*He created the Heavens and the Earth truthfully (justly)*' (the Qur'an 16:03). Here the word 'truthfully' does not accurately translate the original Arabic word in the verse. In fact, the original word indicates that God has created the world justly, to be ordered – and hence, to be apprehended by reason.

We cannot rule out the role of some strict laws that govern the behavior of the world; however, we can realize that the action of such laws are sustained by some supernatural agency that necessarily exists in order to operate these laws. The necessity that such an agency be supernatural is imposed by the fact that if the agency is to be part of our natural world then it has to abide by the same rules of nature, and would therefore be in need of an operator *ad infinitum*.

4. Freedom of the world.

John Polkinghorne believes that 'God has given freedom to the whole world' (Polkinghorne 1990), and he justifies this claim beautifully. According to Polkinghorne, we have *love* and *faithfulness*, as attributes of God, being reflected as *freedom* and *reliability* in the action of the world, which results in the *chance* and *necessity* which we observe. Some people might say that the role of chance subverts the religious claim that there is a purpose at work in the world. But if it is insisted that chance (or so-called happenstance) 'were to be operating within a context of lawful regularity', the significance of 'chance' will be lost. This is why it is better to talk about 'contingency' rather than 'chance'. We might say that the fruitfulness of the world is an interplay between 'contingency and necessity', and this is what the old Mutakallimūn were saying. Contingency is a type of freedom that is given to the world to be as it is.

If we were to claim that God has put certain 'intrinsic properties' into place within nature that allow things to choose their own way to come into action, then this implies that nature has some sort of a decision to make whenever any action takes place. But this would mean that nature has a mind.

The question arises as to how nature might work under the auspices of divine action. Here again we face a God-of-the-gaps, hidden under the notion of 'chance and necessity'. Polkinghorne finds that the 'fruitful interplay between chance and necessity is a reflection of the twin gifts of freedom and reliability which God has given to the world, gifts which are the reflections of his combined nature of love and faithfulness'

(Polkinghorne 1990); but unfortunately, such mystical reflections on this serious ontological matter do not help to resolve the problem of the conflict between God's choice and nature's choice – unless we mean to say that nature's choice may win over the divine Will, in order to explain the existence of natural evil, like cancer. However, it would be difficult to justify why the divine Will might supersede nature's choice; indeed, if we accept this kind of conflict between nature and God we are implicitly assuming the existence of more than one God. It is wisest to admit that this problem is one of the most serious that arises in connection with the reconciliation of science and religion. The challenge posed by this problem comes from the fact that we see nature behaving almost rationally – that is to say, according to reliable laws. Moreover, we know that Nature can adapt itself, and can change the environment in order to develop a balance that favors certain intended goals. This observation has led to the so-called 'Gaia hypothesis' (Lovelock, 2001). An example of this is the adaptation of the environment that took place in the Earth's atmosphere a long time ago, leading to the ratios of oxygen and nitrogen in air that make life possible.

I believe that a better alternative is to assume that nature has no freedom, and that it would be correct to say that it is fully in submission to God. In this context, a sort of divine 'democracy' might be at work instead of a divine tyranny. This submission comes through nature obeying God's orders, through God being the operator and the coordinator of the laws with which nature abides. God, the lawgiver, has designed the world in such a way as to have things be associated with certain properties that characterize what we call the 'nature of things'. Such properties are only at work under the auspices of the divine Will, through the need for co-operation and co-ordination. This leads to the necessity of the world being created anew every moment: the divine action takes place through the operation of given laws. In this way, we can see the true meaning of divine providence.

The question remains: how could the merciful and compassionate God order an evil act of nature to take place? It seems to me that God did not create this world to entertain humans; otherwise, he would not have given it the qualities and laws that would enable natural evil to happen in the first place. Rather, it seems that the challenge put forward for humans in this world is to behave within our given freedoms and capabilities, in order to achieve the goal that we are tasked to achieve. This goal is the destiny behind the whole game of creation, and of the development of humankind and the universe. Theories of cosmology tell us that there could be many worlds empty of any developed life, and that our universe is very accurately fine-tuned to make our existence possible. Therefore, surely humankind is destined to do something of great value that justifies its existence in a world which contains such delicacy and complication.

One final point needs to be made here, and that is to assert that the laws of physics are our rationalizations of how the world reasonably acts. They set out our comprehension of the world; but in no way are these laws necessarily expressing true and actual divine algorithms. These laws are *our* algorithms for the world. The history of the development of the physical concept of nature proves this. Therefore, it would appear that we are far from conceiving how the 'mind of God' works, and that we are far from being able to 'catch God at work', as Einstein was reported to have hoped.

5. Human Freedom.

Human freedom is different from natural freedom, because human freedom implies having one's own intrinsic will; a person can make choices without the need for some external body to dictate them. However, all a person's physical acts need an operator/ coordinator, whether we assume a top-down or a bottom-up causation model.

Human free will is validated by the practical realization of actions. If these actions are not realized in practice, human free will would have no practical value. Since the validation of human will happens through natural effects and actions, it is reasonable to assume that that human will is bound up with the global destiny of the world.

Perhaps the most plausible view in contemporary debates about divine action is that proposed by John Polkinghorne, which assumes that God interacts with the world through the input of 'active information', rather than an input of energy (Polkinghorne, 1990). The idea is intriguing, especially from an Islamic perspective. God always interacts with the world through his words (commands, orders). For example, according to the Qur'an, Jesus was the Word of God revealed to Mary. God is described as the most knowledgeable and experienced, the one who knows all that is hidden. However, Polkinghorne does not appear to suggest any mechanism for actualizing such information transfer. Admittedly, it is difficult to conceive how one can suggest a mechanism of information transfer between two worlds of completely different characters, one physical and the other non-physical. One major problem in this respect is temporality, which is not a common factor between these worlds. In addition, we do not need to insist on the notion of a personal God in order to realize a possible mechanism of information transfer, since the notion of a personal God would imply that God may have a localized body.

6. Mankind and the Universe

According to the Qur'an, the material world has been devised for the purpose of the existence and welfare of humankind. It is there to serve the purpose for which humans have been created. We read, '*See you not that Allah has subjected for you whatsoever in the Heavens and whatsoever in the Earth, and has completed and perfected His graces upon you, both apparent and hidden. Yet of mankind is he who disputes about Allah without knowledge or guidance or illuminating book*' (The Qur'an 31:20). This assertion serves as a reminder of the so-called Anthropic Principle. On the other hand, the Qur'an brings the attention of humans to the delicate and magnificent design of this world, which was executed in accordance with perfect law and order. We read, '*It is He Who made the sun a shining thing and the moon as a light and appointed out its daily positions* (phases) *in order that you know the number of years and the reckoning. Allah did not create this but with the requirement of the truth* (law, order). *He elaborates the Ayat* (signs, revelations, etc.) *for people who have knowledge*' (The Qur'an 10:05).

Summary

To summarize, we can say that matter in the Islamic perspective is considered to be an essential part of the divine creation, and that it is a necessary part of the world in which humankind lives. For this reason, matter, on its own merits, cannot be considered to be sacred; but once it is put to the service of the sacred goal for which mankind was created, it takes on the special quality of being sacred. We may also say that, from an Islamic perspective, the whole universe is considered to be in an entangled state. Space, time, matter and humans are but the creation of Allah by His Will, Who brought up humankind into a state by which the universe becomes intelligible for us within our own capacity; and thus humankind should attempt the challenge of comprehending Allah through the comprehension of his creation. This will then achieve the goal of existence for humans.

Bibliography

Al-Ash'ari, Abu al-Hasan, 1980. *Makalat Al-Islamiyyin*, Arabic text, edited by Helmut Ritter, Wiesbaden: Franzsteis.
Al-Juwayni Abu al-Ma'ali, 1969. *Al-Shamil Fi Usul Addeen*, (in Arabic), Alexandria: Ma'arif Establishment.

Al-Alousi, H.M. 1980. *A Dialogue Between Philosophers and Mutakallimūn*, Beirut: Arab Foundation for Studies, 2nd ed.

Al-Baqillani, 1987. *Kitab Tamheed Al-Awael*, Arabic text edited by Imad al-deen Hayder, Beirut: Al-Kutub Al-Thaqafia Establishment.

Al-Ghazali, 2000. *Tahafut al-Falasifa* (The Incoherence of the Philosophers), translated by Michael Marmura, Utah: Brigham Young University Press.

Altaie, M. B. 1994. 'The Scientific value of Daqīq al-Kalām', *Journal of Islamic Thought and Scientific Creativity*, Vol. 4, No. 2, p.7-18.

—. 2005. 'Time in Islamic Kalām', a paper delivered at the conference 'Einstein, God and Time', Oxford University, September 2005. Unpublished.

—. 2007. 'Islamic Kalam: A possible role in Science and Religion Dialogue', *Annals of the Sergiu Al-George Institute, Bucharest*, XII-XVI, p. 95-109.

Craig, W. 1979. *The Kalām Cosmological Argument*, London and Basingstoke: The Macmillan Press Ltd.

Ibn Hazm, 1983. *Al-Fisal Fi Al-Milal Wal Nihal wal Ahwa'*, Arabic text, Cairo: Al-Khanji Bookshop.

Jammer M. 1974. *The Philosophy of Quantum Theory*, New York: John Wiley.

Lovelock, J. 2001. *Homage to Gaia: The Life of an Independent Scientist*, Oxford: Oxford University Press.

Nasr, S. H. 1978. *An Introduction to Islamic Cosmological Doctrines*, Revised Edition. London: Thames and Hudson.

Pines, S. 1946. *Beitrage zur Islamischen Atomenlebre*, Arabic translation by Abu Reeda. Cairo: Al-Nahdha Bookshop.

Polkinghorne, J. 1990, *God's Action in the World*, J. K. Russell Fellowship Lecture, http://www.polkinghorne.net/action.html (accessed 29 December 2009).

Polkinghorne, J. and Welker, M. 2001. *Faith in the Living God: A Dialogue*. London: SPCK.

Walzer, R. 1970. 'Early Islamic Philosophers', in *The Cambridge History of Late Greek and Early Medieval Philosophy*, ed. A. H. Armstrong. Cambridge: Cambridge University Press.

Wolfson H. 1976. *The Philosophy of the Kalām*. Cambridge Massachusetts: Harvard University Press.

PART III:

THEOLOGICAL PERSPECTIVES ON MATTER

CHAPTER NINE

THE TRIUNE GOD AND THE TRIAD OF MATTER

NIELS HENRIK GREGERSEN,
UNIVERSITY OF COPENHAGEN

Introduction

According to the monotheistic belief-systems, God is creator of the very same world that the sciences are investigating. Advaita Vedanta Hinduism, Judaism, Christianity, and Islam are monotheistic religions that share a fundamentally positive attitude to material world, in the sense that they commonly assume that Brahman, Yahweh, God the Father, Allah is ultimately responsible for the basic cosmic structures. 'God saw everything that he had made, and indeed, it was very good', as it is said in the Hebrew Bible (Gen 1:31).

The major monotheistic religions, however, each of them divided into many distinctive schools of thought, do not concur on the relation between the ultimate reality of God and the material world. In what follows, I propose a Christian view of the material world, which sees God as being intimately present in the core of physical matter – without conflating God and creation, but also without separating God from the material world. More specifically, I shall argue that the idea of a Triune God – Father, Son, and Holy Spirit – offers unique resources for developing an ontology which is congenial to current scientific concepts of physical matter as constituting a world of *mass, energy, and information*. Accordingly, this proposal stands or falls with the assumption that the concept of information should be accorded an ontological role in our current scientific concept of matter (as is argued in Davies and Gregersen 2010). As such, the proposal is not only fallible in a general sense, but is also open for a falsification in light of future scientific research. Needless to say, however, the proposed Trinitarian view of the material world does not in itself have any status as a scientific theory. Hence my claim is not that any scientist who is committed to a broader view of the material *must*, by implication, also embrace a Trinitarian view of the interrelation between mass, energy, and information. Rather it is a metaphysical proposal developed under the specific constraints that the theological proposal must be congenial with recent developments in scientific understandings of matter. In short, I am presenting a piece of a *theology of nature* (coming from theology to physics and biology) rather than developing an argument exclusively from below, in the otherwise great tradition of *natural theology*.

The Demise of the Matter Myth

In the epoch between 1700 and 1900 materialism was a position favored by the physical sciences; a theological interpretation of God as the ultimate lawgiver could be added to the picture, but the corpuscular theory of matter was sustained. This inherited concept of matter, however, became obsolete with the energy-matter unities of relativity theory. This was perceived early on by Bertrand Russell:

> Matter, for common sense, is something which persists in time and moves in space. But for modern relativity physics this view is no longer tenable. A piece of matter has become, not a persistent thing with varying states, but a system of inter-related events. The old solidity is gone, and with it the characteristics that, to the materialist, made matter seem more real than fleeting thoughts (Russell 1961, p. 241).

Even more revolutionary for the concept of matter was quantum mechanics, which gave up the idea of material entities having a simple definable state at the ultimate level of matter. The uncertainty principle of quantum mechanics shows that particles emerge out of, and perish into, a field of subatomic events with an ontological status that defies description in terms of simple location or duration. The stochastic nature of the decoherence of quantum events into classical events, and the continuous entanglement of otherwise distant events, give evidence that there is no scientific basis left for the common sense view of matter. Matter has become an ungraspable concept. Atoms are not un-dividable entities, as the old etymology of *a-tomos* (Greek for un-divided) suggests. Nor are atoms, according to entanglement, always separated from one another according to the scheme of causes and effects, ruled by necessitarian laws.

Discussing the consequences of relativity theory and quantum mechanics, the philosopher Norwood Russell Hanson even spoke about the 'de-materialization' of the scientific concept of matter: 'Matter has been dematerialized, not just as a concept of the philosophically real, but now as an idea of modern physics. ... The things which for Newton typified matter – e.g., an exactly determinable state, a point shape, absolute solidity – these are now the properties electrons do not, because theoretically they cannot, have' (Hanson 1962, p. 34). Hanson's point, of course, is not that physical events do not have a material basis, but that the concept of matter is undergoing serious revisions. The visibility, indivisibility, and locality of old-style materialism have gone.

In fact, this development had already precursors in the development of thermodynamics in nineteenth century physics. In his *Remarks on the Forces of Inorganic Nature* (first published in 1842), the German natural philosopher Julius Robert Mayers formulated a principle that pointed forwards to a fundamental change in the scientific concept of matter. The essential property of force or energy, according to Mayer, consists of 'the unity of its indestructability and convertability' (Mayer 1980, p. 70). A little later, in 1851, the English physicist William Thomson (later Lord Kelvin) intimated: 'I believe the tendency in the material world is for motion to become diffused, and that as a whole the reversion of concentration is gradually going on' (Thomson 1980, p. 85).

Thus the intuitions were formed that were going to be formulated in the first and second laws of thermodynamics. The first law of thermodynamics states that energy is conserved when it is put into work and converted into heat. Thereby heat appeared to be a general property of matter. In 1865 Rudolph Clausius then formulated the second law of thermodynamics, saying that energy exchanges are irreversible. In a closed system, a portion of energy converted into work dissipates and loses its force to do the same work twice. Thus the energy is at once a constant feature of matter, and a more and more inefficient capacity of matter, as time goes on. Understanding the universe as a closed system by assuming the first law of thermodynamics, the law of entropy predicts the bleak perspective that the universe is going be less and less capable of producing the heat necessary for living organisms to survive.

With Albert Einstein's general theory of relativity, the concept of energy was further generalized by understanding mass and energy as quantitatively equivalent. In a vacuum, energy (E) is numerically equal to the product of mass (m) and the speed of light squared: $E = mc^2$. At a closer look, this famous formula can have two different philosophical interpretations. It can mean that 'mass' and 'energy' are two equal properties of an underlying material system, or it can be taken to mean that energy and mass constitute the same stuff, which then appears with different emphasis in different systems. In some systems the mass-aspect of matter dominates, while at other places matter takes the form of a field. In Einstein's and Infeld's *The Evolution of Physics*, the latter view is expressed as follows: 'Matter is where the concentration of energy is great, field where the concentration of energy is small' (Einstein and Infeld 1938, p. 242; Flores 2004, pp. 4-6). This distinction between matter and field view especially highlights the fact that most matter is not visible in any form, but can only be evidenced indirectly by its gravitational force. (The difficult issue of the status of 'dark matter', emitting no light but exerting a gravitational effect, arises in this context.) Matter is no longer what it used to be.

Also, biology posed new questions to materialism by bringing *information* into the centre of attention. Despite all claims of causal reduction, physics (not only classical, but also modern) has failed to explain basic features of biological evolution. Even if some chemical compounds are formed under certain circumstances due to atomic affinities fully explicable by physical laws, there exists no law for the sequences of the DNA macromolecules. Thus, genomes are arbitrary relative to underlying chemical affinities. Genomes are as they are due to contingent historical circumstances. If DNA-sequences are causally efficient instructors by virtue of their informational structure, information can no longer be left out of a comprehensive picture of what drives nature. For what is causally effective must be given status as something real. Eventually, information codetermines how organisms make use of their available energy budgets.

Information at the Core of Matter: From Qubits to DNA

Skeptics may still want to question whether 'information' is as seminal to nature as are the mass-and-energy aspects of physical matter. For whereas we have physical measure for mass and energy in terms of the units of *grams* and *joules*, we do not seem have to a corresponding measurable unit for information. One may therefore suspect that 'information' is just a metaphor that we use as a shorthand for various different purposes, as when we speak about information technology, about instructions or recipes, or simply about anything that some way or another makes sense to us.

As we shall see below, several distinct meanings of the term information are in need of being sorted out, at least in broad terms: information as digits (Shannon information), information as patterns (Aristotelian information), and information as meaning (semantic information). However, there are two reasons for giving information a central role in a scientifically informed ontology (that is, a view about what really matters and makes a causal difference in our world). First, information could be said to be at the rock bottom of physical reality in terms of the generative capacities of quantum processes. And second, informational structures play an undeniable causal role in material constellations, as we see in, for example, the physical phenomenon of resonance, or in biological systems such as DNA sequences. No bridge and no skyscraper can be constructed successfully without paying due attention to the phenomenon of resonance, and no evolutionary theory can work without attending to the instructional powers of DNA sequences. Just as *informational events* are quintessential at the bottom level of quantum reality, so are *informational structures* the driving forces for the historical unfolding of physical reality.

That information is seminal to quantum mechanics is a main thesis in Seth Lloyd's *Programming the Universe* (2006). According to Lloyd, the universe is *not* like a digital computer (as suggested by Stephen Wolfram), but is a quantum computer, whose fundamental bits are qubits (quantum bits). Why use the term information about quantum events in the cosmos? Because each and every quantum event not only *does* something on the basis of the immediate situation of the universe (that is, performs an energy transaction), but by its occurrence it also *instructs* (informationally) the situation immediately following, in which other quantum events are going to occur. Since the relation between two quantum events A and B is not unilaterally predetermined from A (even though the over-all relations between As and Bs take place within the statistical limits of quantum mechanics), qubits do not act, and do not make instructions, in the same manner as a digital computer that runs its programs on a classical hardware system. Qubits do not behave like binary digits, for which in each computational step there are only two possibilities ('0' or '1'), one of which is predetermined by the software program ('when x, do 0: when y, do 1'). Rather, the behavior of qubits affords many possible outcomes. In the words of Seth Lloyd, 'Every molecule, atoms, and elementary particles register bits of information. Every interaction between those pieces of the universe processes that information by altering those bits. That is, the universe computes' (Lloyd 2006, p. 3).

Quantum events thus *produce differences*, and these differences both make up the *status quo* of the universe at any given time, and inform the subsequent cosmic situation. As argued by Lloyd, 'information and energy play complementary roles in the universe. Energy makes physical systems do things. Information tells them what to do'. According to Lloyd, 'the primary actor in the physical history of the universe is information' (Lloyd 2006, p. 40).

Such instructional powers of information come even more to the fore in the world of biology. Metaphorically speaking, one might say that the language of information in quantum mechanics speaks in terms of distinctive *cuts*. In each moment *this* happens, *not* that: 'cut, cut!'. But the world of the living is constituted by informational process that create resonances: 'build up, build up!' is the language of biological systems. In a programmatic article, 'The Concept of Information in Biology', the late John Maynard Smith argued that information is a quintessential concept in biology because DNA – like culture – is concerned with the storage and transmission of information. Moreover, he argues that there is a fundamental difference between the *genes*, which are 'codes' that instruct the formation of specific proteins, and the *proteins* that are coded for by the genes. Even if proteins themselves often have a strong causal effect as inducers or repressors of genes, their causal role is arbitrary in relation to the genes. Thus there is no necessity about which inducers regulate which genes; it is a question of happenstance, of *gratuité* (to use Jacques Monod's term). By contrast, genes code for something very specific to take place, such as the production of eyes. Hence genes are not only informational in the sense that they *store information* about a biological past; genes are also informational in the sense of *providing instructions* for subsequent adaptive purposes. In this sense, argues Maynard Smith, genes are not only of central importance for molecular biology, but also for developmental biology. The informational structures of genes are thus intentional, in the sense of seeking a desirable outcome – though not in the mental sense of meaning information (Maynard Smith 2000).

The story of the increasing causal capacity of informational systems could easily be extended into the human world of information technology. Here, machines are constructed for purposes that are felt to be of importance for human users, either as instruments for achieving purposes (say, making helpful calculations), or as ends in themselves (say, television entertainment). With the advent of the computer sciences, the idea of matter as an assembly of isolated bricks has now lost its plausibility as a common sense idea about reality. Also, virtual reality may have a causal impact. At least from a human perspective, patterns of information matter more than bits of information; and information which is felt to be 'meaningful' or important to human observers matters more than the objective patterns of nature in themselves. Ultimately information matters, since informational structures have causal efficacy.

As a result, our concept of matter may not have been simply demolished, but it certainly has been transformed and enlarged so that (1) the *substance* of matter (that is, its basis in material stuff such as particles), (2) the *energy* of matter (the potential and actual changeability of physical matter), and (3) the *organization* of matter (its capacity for pattern formation) constitute three relatively independent, though inseparable, aspects of matter, which cannot be fully reduced to one another (c.f. Zemann, 1990, p. 695: 'Der Begriff der Materie hat drei Hauptaspekte, die untereinander untrennbar, aber relativ selbständig sind: der stoffliche (mit Beziehung zum Substrat), der energetische (mit Beziehung zur Bewegung) und der informatorische (mit Beziehung zur Struktur und Organisiertheit)'). Especially in biology, the notion of information is important, since information is used to determine how to use the energy budget available for each organism, and for interactive group behaviour. As argued by Christopher Langton, '[i]n living systems, information processing has somehow gained the upper hand over the dynamics of energy that dominates the behavior of most non-living systems (Langton 2006, pp. 42-3).

Three Aspects of Information that Matter

The philosophical question then arises, What is meant by information? John Puddefoot has helpfully distinguished between three types of information (Puddefoot 1996). The first is *counting information*, or the mathematical concept of information as defined by Claude Shannon. Here 'information' means the minimal information content, expressed in bits (*binary digits*), of any state or event: '1' or '0.' In order, for example, to know which of sixteen people have won a car in a lottery, one would expect to need sixteen bits of information, namely the fifteen losers (symbolized by a '0') and the one winner (symbolized by a '1'). However, if one began to ask intelligently ('is the winner in this or that group of eight', and so on), one would only need four computational steps ($\log_2 16 = 4$) (Weaver 1949, p. 100f). Here 'information' means mathematically compressible information. This counting concept lies at the heart of Computational Complexity, including the idea of Artificial Life.

Different from this is what Puddefoot calls *shaping information*, which is the form or pattern of existing things. Here the interest lies in morphology, the study of forms. Shaping information may derive either from internal sources (such as a zygote), or from external constraints that cause something to have a definite pattern in relation to its environment (such as inducers or repressors). This aspect of information relates to what we above called constellations, or patterns. And, as we have seen, these play an important role in evolutionary processes in living systems, and even in the formation of snow crystals and other semi-stable non-living systems.

The third type of information is what we refer to in daily parlance: coming to know something of importance – to an observer. In *meaning information,* information means something *to someone* in a given context. A facial expression, for instance, can convey kindness or aggression, the awareness of which again makes a vast difference in a social context, with undeniable causal consequences.

I do not think that leaps from one type of information to another are possible. It seems, however, that the concept of *shaping information* is basic in relation to the two other forms of information. In physics we found examples of this in Seth Lloyd's computational view of quantum processes as well as in DNA-instructions of biological systems. Digital counting information seeks to model or simplify shaping information, whereas meaning information seeks to interpret shaping information for a specific purpose.

The new role of information (and the corresponding idea of a 'local' or domain-specific physics) offers, as far as I can see, two new possibilities for a theological understanding of nature. First, there is in the world of nature a richness and heterogeneity – a natural complexity – that discourages the old-style reductionist claim that we and our fellow creatures are 'nothing but molecules' (J. Monod), or 'nothing but extended phenotypes of self-replicating DNA' (R. Dawkins). The reductionist idea that nature is uniform and can be comprehended from the perspective of one single overarching scientific theory (molecular or genetic) presupposes that information can be left out of the picture of reality, to be replaced by mechanical processes of energy transactions. But since organization is what ultimately matters in biology (*pace* Maynard Smith), physicalist versions of reductionism are dead ends for the purpose of understanding biological life (though they may explain the origin of life). We should not be bashing reductionism altogether, however. Reductionism can and should be pursued as a *method* for the many types of investigation which can be done while isolating the parts from the wider system. The neurosciences, for example, should try to reduce general neuronal patters to more specialized and localizable parts of the brain. But reductionism can no longer be presented as the only science game in town. In order to achieve long-term progress, the neurosciences must also attend to resonances emerging between different parts of the neural system, and include the importance of the environment which constantly feeds the brain (and the wider neural system) with new inputs of potential importance for the organism as a whole.

By implication, no neuronal theory can claim to be complete, unless it is able to explain the physiological effects of viewing, say, a horror movie or a thriller. This again presupposes technological devices such as movie screens as well as cultural learning patterns, without which one could not understand a crucial phenomenon as well-known as Alfred Hitchcock's 'suspense'. For states of suspense presuppose information about what is *absent* but *anticipated.* Curiously enough, also the type of information which is missing can be causally effective. Think of everyday situations where we say hello, but are not greeted back. Or think of the painful situation when the young man proposes to a girl, but receives no answer. And *that* is exactly the answer! A suspension, and an absent response, are thus examples of *missing information*, which functions as the information that matters, exactly by virtue of being absent (c.f. Deacon 2010).

Non-reductive Societies:
From divine Trinity to material Triads

So far I have focused mostly on the role of information, since informational structures of matter tend to go unnoticed in the physical sciences, and are often relegated to the more soft scientific disciplines such as psychology and social science, if not to mere 'folk psychology'. But information is 'the difference that makes a difference', as famously expressed by Gregory Bateson. Hence it should be accorded an ontological status alongside mass and energy in a current scientific view of matter and the material world.

Coming so far it is difficult not to see some formal similarities between the three irreducible aspect of matter, and the Christian doctrine of the Trinity of Father (the substance), Son (form), and Holy Spirit (energy). Let me therefore present a theological thought experiment concerning the ontological relationship between God and the material world. I hereby hope to show not only why God and physical reality are one and united, but also how God differs from the world of matter.

Let me begin by clarifying the meaning of Trinitarian faith. For just as scientists hesitate to speak about 'matter' in general terms, but prefer to discuss matter under specific conditions, so theologians avoid speaking in general terms about 'divinity' or 'spirituality.' Why not be as specific as possible, also in theological matters? In the fourth century, the Cappadocian Fathers (Basil the Great, Gregory of Nyssa, and Gregory of Nazianzus) interpreted the Christian confession of God as Father, Son, and Holy Spirit as follows.

(1) The three divine persons making up the divine being (Greek *ousia*) are seen as a living communion (*koinonia*), which is realized only in the eternal reciprocal interactions (later called *perichoresis*) between the three

divine persons (*hypostaseis*). The Father, the Son/Logos, and the Spirit are thus inseparable from one another in the one community of God's being. Yet they are not reducible to one another, but remain as three distinct persons or poles in the divine communion.

(2) More specifically, God is conceived of as a living community, in which the Father eternally generates (makes room for) the otherness of the Son, while the 'life-giving Spirit' proceeds from the Father in accordance with the Son or Logos. At the same time, the Father could not be Father without giving birth to the Son, who makes the Father a father (or mother), just as the relation of love between the Father and the Son would not be fulfilled without the Spirit as the living realization of personal love between them. The divine 'Persons' are thus co-constituted by one another.

(3) While the Father is the eternal source or 'fountainhead' of the Trinity, the eternal Son and the eternal Spirit are the Divine Persons who issue forth from the pure, ineffable divinity and in the act of creation establish the world of creation. The Son and the Spirit are the 'two hands of the Father,' as Irenaeus had already put it around the year 200. But neither would the Father be the creator, and thus not 'the Father of all' (Eph. 4:6), without the divine Logos who incessantly informs the world of creation as its creative Pattern in constant communication with the Spirit, who continues to energize and transform the world.

And now to our *Gedankenexperiment*. How may we envisage the relation between God and the material world if we assume (premise 1) that God is an eternal Trinitarian community, and if we assume (premise 2) that the three aspects of matter – substance, energy, and organization – are inseparable but irreducible aspects of our universe?

One possibility is to think of God and matter as two names for the same underlying reality. This is the position of *pantheism*. Pantheism remains a possibility in the wake of complexity theory. A version was formulated by Samuel Alexander (one of the main emergentists in the 1920s) and has recently been rearticulated by Harold Morowitz, for whom the laws of nature are identical with the immanent God. But eventually the rules of selection give rise to the emergence of *homo sapiens*. At some point in time, consciousness entered into the picture of nature, and nature was thereby transcended for the first time. By means of consciousness, we can project different futures and become responsible for our own actions. 'We, *Homo sapiens*, are the transcendence of God' (Morowitz 2002, p. 196). God, by contrast, is purely immanent.

The question is whether the identification of God with the laws of nature comes even remotely close to what is usually meant by 'God' in religious language. In Morowitz's portrayal, God is dumb, hence without will, without knowledge, without feeling. God would not even be the cause of the world in the full sense of the word, since the laws of nature presuppose some initial conditions to have existed prior to the laws themselves. Matter would exist independently of God as the subvenient base of the supervenient divine spirit, which then emerges late in the history of the cosmos. Moreover, any physical, biological, or historical event would be an immediate expression of God's underlying nature. The Holocaust would be a divine revelation to exactly the same extent as phenomena such as faith, hope, and love. Although I admit the logical possibility of pantheism, and even though I also acknowledge the intellectual economy of the idea, I do not find it an attractive hypothesis, since it cannot capture the phenomena of religious, moral, or aesthetic experience.

I suggest another route. First, God and matter are so deeply interconnected that one could say, with Anselm of Canterbury: 'Where God is not, there is nothing' (*Monologion* 14). This is to say that wherever something exists, God is radically present in it (that is, present in its *radix*, its root). Furthermore, if one takes a position of a minimal naturalism (saying that all-that-exists within the world of creation is materially constituted), then God, by implication, is present in matter itself. This is what is suggested in the Old Testament (Jer 23:24; Ps 139) as well as in the New Testament (Acts 17:28). In the writings of Martin Luther, we find perhaps the most striking affirmation of the unity between God and material beings (see Gregersen 2005). In discussing how God can 'really' be present in the bread and wine of the Eucharist, Luther takes a strong view in favor of God's omnipresence in the world of matter: 'For it is God who, through his almighty power and right hand, creates, works in and contains all things. ... Therefore He must fully exist in each individual creature, in their intimacy as well as around them, through and through, below and above, in front and behind, since nothing can be more present or intimate in all creatures than God with His power' (Luther 1964 vol 23, p. 133f.: translation by the author. c.f. Luther 1961 vol. 37, p. 57f.). Luther here expresses a theological ontology, but does not develop a natural theology. The claim that God is present in all material parts of the universe does not mean that God is revealed in all things. On the contrary, Luther holds the view that the God who is everywhere in the cosmos is usually hidden to us, but made graspable in God's own Word: Jesus Christ. For God is in all things without being identical with any thing. God is omnipresent without being 'omnimanifest.'

With the new concept of matter enriched by information theory, I believe that this theological position can be made more precise. *On the*

Trinitarian view, God is not identical with matter, nor is the Father identical with the stuff of matter (mass). For if this were the case, the identity of God as an eternal communion of love would not be safeguarded; all things that happen in our world would immediately be identical with God's own being-and-willing. Rather, the claim is that God in creation makes room for the otherness of matter, similar to the manner in which God, within the internal Trinitarian life, makes room for the otherness of the 'Son'. The Son is thus the principle of otherness within God's own life, and the divine creativity is thus said to take place through the Son. This is an important insight of Hegel's rediscovery of the Christian doctrine of the Trinity (see Pannenberg 1991, pp. 42–49). Pannenberg criticizes Hegel for making creation a logically necessary act of God, and he adds that the act of creation is based not only on the Father's eternal birthing of the Son, but also on the Son's free self-distinction from the Father – a distinction, however, which never leads to a separation due to the work of the Spirit.

The term 'Son' is derived from the Jesus history. However, we should bear in mind that the term 'Son' is a familial metaphor for the manner in which the eternal Logos expresses the mind of God in an uncompromising manner, while remaining distinct from the Father (the one and only Source of all being) by having his own will and self-awareness. In this sense, the eternal Father makes room for the eternal Logos, and this Logos ('in God') is the principle of otherness, who freely distinguishes itself from the Father in order to be God's 'hand' in creation. This is why it can be said in the New Testament that all things are 'from' the Father (who is the 'whence' of physical stuff), but that all things have come into being 'though' the Son (1 Cor. 8:6). For the Logos is the Creative Pattern for all that is 'logical' in the world of creation. *The Son/Logos is the free, outward expression of the inner mind of God the Father, and as such the Creative Matrix that releases pattern formation in the midst of creation during evolution.*

It is helpful to bear in mind the full range of the Logos-Christology. Indeed, from the perspective of Trinitarian faith, the Logos has to do with the informational aspects of all that is. Christ does not enter history with the man Jesus, in whom the Word became incarnate, but the Logos or the eternal Son was ever-present in God (John 1:1), and united himself not just with a single human being but with the 'flesh' of materiality (John 1:14). The Logos is God's readiness to let the world be, and to let Godself become embodied in the material world. Hence, incarnation is *deep*, not shallow or bounded to a particular time or place. The Logos is the creative Pattern of the universe, the informational Matrix from which all patterns arose, arise, and will continue to emerge throughout history (John 1:3). *As such, the Logos is not identical with matter's informational capacities, but the wellspring of order in the universe.*

What, then, is the corresponding role of the Holy Spirit? The role of the Holy Spirit is precisely to energize the world of creation. The first thing said about the Spirit in the Nicene Creed is that it is the vivifying spirit, 'the giver of life', who proceeds from the Father (and the Son). Only then is it said that the Spirit also spoke in historical time to the prophets, and thus has been guiding humankind through historical time. Just as Christ has a cosmic function as co-creator before and after the time of Jesus, so the Spirit has a function in nature before and apart from the more spiritual aspects of human existence. Therefore, we again find that the Spirit is the Giver of life, but not itself a biological being with DNA, and so on. And yet, *as the transcendent source of life, the giver is present in the gifts of biological existence, though without being identical with specific life-forms.* If this were the case the divine Spirit would die with the end of biological existence. But the Spirit is the eternal energizer of divine life, who is present in all vital functions of all living beings, and who continues to release the ambience of mutuality in the world.

Now the triadic structure of divine life also shows its inner logic. If God the Father had only one 'hand', be it the Son or the Spirit, the complex world could not come into being. Imagine if only the Logos had been sent into the world, but not the Spirit: there would be structure everywhere, but no life and no movement. And imagine if only the Spirit had come into the world, but not the Logos: there would be chaotic processes everywhere, but no structure anywhere from which the world of organized complexity could grow forth.

It is exactly in the interplay between Logos and Energy, that creativity unfolds itself and emergence appears – between pure order and pure randomness.

Bibliography

Davies, P. and Gregersen, N. H. (eds.) 2010. *Information and the Nature of Reality: From Physics to Metaphysics.* Cambridge: Cambridge University Press.

Deacon, T. 2010. 'What is missing from theories of information?' in Paul Davies and Niels Henrik Gregersen (eds.), *Information and the Nature of Reality: From Physics to Metaphysics.* Cambridge: Cambridge University Press.

Einstein, A and Infeld, L. 1938. *The Evolution of Physics*. New York: Simon and Shuster

Flores, F. 2004. 'The Equivalence of Mass and Energy', *Stanford Encyclopedia of Philosophy*, accessed August 2, 2006 at http://plato.stanford.edu/entries/equivME.

Gregersen, N. H. 2005. '*Unio creatoris et creaturae*: Martin Luther's Trinitarian View of Creation', in Else Marie Wiberg Pedersen and Johannes Nissen (eds.), *Cracks in the Wall: Essays on Spirituality, Ecumenicity and Ethics*, pp. 43-58. Frankfurt am Main: Peter Lang Verlag.

Hanson, N. R. 1962. 'The Dematerialization of Matter', *Philosophy of Science*, pp. 27-38.

Langton, C. G. 2006. 'Life at the Edge of Chaos', in C.G. Langton et al. (eds.), *Artificial Life II: Proceedings of the Workshop on Artificial Life held February, 1990, in Santa Fe, New Mexico* , pp.41-92. Santa Fe, NM: Santa Fe Institute Studies in the Sciences of Complexity.

Lloyd, S. 2006. *Programming the Universe. A Quantum Computer Scientist Takes on the Cosmos*. New York: Alfred A Knopf.

Luther, M. 1961. *Luther's Works vol. 37*. Philadelphia: Fortress Press.

—. 1964. 'Das diese Wort Christi "Das is mein Leib" noch feststehen, wider die Schwärmgeister', in *D. Martin Luthers Werke vol. 23*. Weimar: Hermann Böhlaus.

Mayer, J. R. 1980. 'Remarks on the Forces of Inorganic Nature', in Noel G. Coley and Vance M.D. Hall (eds.) *Darwin to Einstein: Primary Sources on Science & Belief*, pp. 68-73. Harlow, Essex: Longman.

Maynard Smith, J. 2000. 'The Concept of Information in Biology', *Philosophy of Science* 67:2 pp. 177-194.

Morowitz, H. J. 2002. *The Emergence of Everything: How the World Became Complex*. New York: Oxford University Press.

Pannenberg, W. 1991. *Systematische Theologie, Band 2*. Göttingen: Vandenhoeck & Ruprecht.

Puddefoot, John C. 1996. 'Information Theory, Biology, and Christology', in Mark Richardson and Wesley J. Wildman (eds.), *Religion and Science: History, Method, Dialogue*, pp. 301-320. New York: Routledge.

Russell, B. 1961. 'Introduction to *A History of Materialism*, by F.A. Lange (1925)', in Robert Egner and Leister E. Dennon (eds.), *The Basic Writings of Bertrand Russell 1903-1959*, pp. 237-245. New York: Simon & Shuster.

Thomson, W. 1980. 'On the Dynamic Theory of Heat', in Noel G. Coley and Vance M.D. Hall (eds.), *Darwin to Einstein: Primary Sources on Science and Belief*, pp. 84-86. Harlow, Essex: Longman.

Weaver, W. 1949. 'Recent Contributions to the Mathematical Theory of Communication,' in Claude E. Shannon and Warren Weaver (eds.), *The Mathematical Theory of Communication*, pp. 94-117. Urbana: The University of Illinois Press.

Zemann, J. 1990. 'Energie', in Hans Jörg Sandkühler et al. (eds.) *Europäische Enzyklopädie zu Philosophie und Wissenschaften*, Bd. 1, pp. 694-696. Hamburg: Felix Meiner.

CHAPTER TEN

A RESPONSE TO NIELS HENRIK GREGERSEN'S
GOD, MATTER AND INFORMATION

KENNETH WILSON,
THE QUEEN'S FOUNDATION, BIRMINGHAM

This is an intriguing, encouraging, and in many ways thrilling paper, because it points to the dynamic nature of the universe in which both God and we are actually participant while at the same time being distinctive and independent, and not being separate. In such a rich and complex paper it is only possible to draw attention to a few aspects which particularly grabbed my attention, and to raise a few of the questions which come to mind.

First, as opposed to much theological enquiry which is concerned to come to terms with – even to answer – the question of how God and the world, the divine and the human, the spiritual and material, can interact with one another, Gregersen, in the light of a discussion of developments in contemporary physics and biology, simply sets about the task of uncovering and illuminating the relationship. Thus whereas some traditions have in the past argued – or at least assumed – that a relationship could only be established miraculously through God's gracious choice and direct intervention, the position here is that the very nature of the world and of God are such that it is unimaginable, and almost inconceivable, that they could be independently explained and then a link sought between them.

The former impossible anxiety came from the Greek and Hebrew traditions. Gregersen employs the intimacy of the Stoic conceptual scheme and its account of nature because there is evidence, he believes, to support the view that Stoic thinking lies behind the prologue of Fourth Gospel; but he is careful to say that he does not believe that it is exclusively the product of Stoicism, and he looks for a balance of the Platonic and the

Stoic traditions if we are to interpret Christian experience with the appropriate, vital dynamic.

But even to this limited extent one must be careful; there are dangers in the world view of Stoicism – and of course in Platonism, but that is another story – which are inimical to the essential generosities of human community and the divine relationship with the world as conceived and developed within Christianity. For example, Stoicism assumes a law of nature which is invariant: even if it is possibly dynamic, it is bereft of emotion or sympathy. To put it in the term of Christian theology, it lacks grace. Hence for the Stoic, admirable human behavior, rightly conducted, will be in accordance with the implacable events of the natural world and leave no room for the influence of human desire or the normal virtues. Where does the debate about ethics, the practice of the virtues, fit into Gregersen's world of matter?

Second, I liked very much what might be called the 'natural' account of everything and the desire to tell it in one coherent, illuminating, developing and creative story. What is more, it is not, as Gregersen sees it, an attempt to impose a 'new' theology upon our contemporary understanding of physics, but rather a development of the tradition in the light of current experience. The story which has begun is continued. Thus, when matter is unpacked as mass, energy and information in its many forms, it is revealed to offer a lively conceptual model which can not only be used to inform our understanding of the world physically; it opens up a helpful understanding of God as Trinity and introduces us to a shared approach to God's relationship to the world. In particular, Gregersen draws on John Puddefoot's distinction between three types of information: counting information, which may be carried out in an infinite number of ways according to the purpose in mind: shaping information, which pertains to morphology, the pattern or shape of existing things (this is fundamental to the other two); and meaning information, which shows the value to us of the information in terms of what it means to us. This is neat indeed, and actually very exciting.

I am reminded of Wittgenstein's work on language, and of the way in which he wrestles with the question of its reference. He distinguished between 'saying' and 'showing'. Language which 'says' something is language with which the speaker is involved, the very saying of which may make a difference to the speaker and to the world of which the speaker is an aspect or dimension. (See Gregersen's reference to qubits (quantum bits) and the possibility of many outcomes, on p. 107 above.) Language which 'shows' something is language which purports to have a certain coherence that enables it to refer to something beyond itself: to have, if you like, a metaphysical significance. There is much debate focused particularly on the writing of Cora Diamond as to whether Wittgenstein ever held a metaphysical view. Perhaps his 'showing' only had the objectivity associated with revealing the nature of the 'language game', or what he called its grammar. Perhaps we cannot get beyond the language. Given, of course, that in history we have run up so many dead ends and back alleys in our science (and in our accounts of our human experience more generally), it is well to have a large number of ways of talking, a large number of languages open to us, because the use of any one of them individually or in combination is likely to throw up possibilities of many other explanations, potential experiences and experiments. If this is so, what is the basis on which it could be argued that Gregersen's thesis is true? What does 'true' mean here?

Third, I was intrigued in the light of this point that Gregersen should insist that his thesis (that the Logos is the informational resource of the universe) is scientifically falsifiable, in the event that the concept of information could be reduced to properties of mass or energy transactions. I am not clear that this is the case, since it seems to me that, given the conceptual scheme which he adopts, it is a matter of definition, and not as a result of empirical enquiry, that the Logos is so identified. On the other hand, I am not sure that this matters, since the language game he (rightly) employs makes no claim to be a natural theology based on science. 'Rather', he states, 'it is a metaphysical proposal developed under the specific constraints that the theological proposal must be congenial with recent developments in scientific understandings of matter.' Actually, that is what matters. Each dimension of explanation, scientific and theological, can be creatively debated with the other to their mutual benefit: they will not be disproved, but to the extent that we find them useful and exciting they will continue to be debated to our profit.

Fourthly, the 'wholeness' of our experience of the world which is offered here bears testimony to the fact that we cannot describe it completely, but we can develop styles of thinking and speaking which enable us to glimpse it and to engage with it, thus thereby continuing to engage in the unending task of understanding it. The process of attending to the evidence and trying to understand our experience is mutually creative of ourselves, one another and the world. There will never be rules which cover all cases because the truth cannot be reduced to one system, one science, or one language. As Alistair MacIntyre so eloquently points out, in his seminal books *After Virtue* and *Whose Justice? Which Rationality?* (MacIntyre 1984, 1988), ethics cannot be exclusively governed by obedience to laws, because we are always required to make judgments about new experiences which extend our experience and add to the quota of information which, on reflection, may change our moral perspective. Neither can we reduce our ethical judgments to those which are right in our own eyes, because the traditions to which we are attached have a life of their own through our participation in the various communities of which we are members. In like manner, the interactions of the levels of information with which we are engaged are not reducible to one, because they have a life of their own; but at the same time no one type can be used to make sense of our experience without reference to the others.

Here indeed are ways of approaching the wholeness of the nature of God as Trinity, along the lines of those which have been explored in the Eastern tradition of *perichoresis* attributed to Cyril of Alexandria in the 5th century, developed by St John of Damascus, and found earlier in the Cappadocian Fathers. We cannot grasp Him in any old fashioned sense of attributing matter to Him, but we can explore and grow in relation to Him by trying to understand our experience of ourselves, the world and one another, as God's creation. I am grateful for the introduction to these language games, which I find very helpful.

Bibliography

MacIntyre, A. 1984. *After Virtue*. Notre Dame, Indiana, Notre Dame University Press

—. 1988. *Whose Justice? Which Rationality?* London Duckworth.

CHAPTER ELEVEN

DIVINE GRACE AND THE CREATED ORDER IN THE HISTORY OF CATHOLIC THEOLOGY

HILARY C. MARTIN

Introduction

The theology of the relationship of nature and divine grace has undergone a series of emphases and changes through two centuries of Christian reflection upon the interplay of humanity and the order which transcends the natural world. It is a subject which addresses some fundamental questions concerning the human being as part of earthly reality and yet formed in the image of God. If any model of creation is theo-centric, there remains the question as to whether we can understand sacredness as incorporated into the natural order or whether the unseen, or spiritual, exists apart from the world. The question of these two positions – intrinsicism and extrinsicism – has been a contentious issue in the history of Catholic thought. The sanctity of human life is affirmed by the Church (Catechism of the Catholic Church 1994, para. 2258, p. 486), but what precisely is it that calls for the use of the word 'sanctity'? What exists within the human condition that distinguishes it from the remainder of the natural world besides language, a highly developed capacity for reason and the ability to apply thought processes outside the confines of time and space – all attributes that may be accounted for in a purely psycho-physical sense? This paper focuses upon divine grace and the human condition, but it has implications for the whole of the natural order of which humanity is an integral part. Instrumental in the way in which nature is perceived is the purpose and end towards which the natural order is progressing in time and, from a Christian viewpoint, it is understood as moving towards a God-ordained fulfilment. In this context, it is difficult to imagine divine grace as totally extrinsic to human nature and, indeed, to nature as a whole. If God is seen as Creator, then re-creation is a logical possibility within that context.

The early church, eager to steer Christianity away from the animism of pagan religions and to emphasise the 'otherness' of God, encouraged a view of God as separate from the world. The Scriptures, however, reveal a God of intimate involvement in both nature and the affairs of men and, throughout Christian history, the questions surrounding God's immanence and transcendence have been a source of debate. In seeking to understand reality in both a physical and metaphysical sense, different perspectives have been held on of the relationship of matter and form, body and soul, the natural and the supernatural. In Greek philosophy, Aristotle's unified idea of reality, where matter and form, the material and the non-material, are interdependent, stands in contrast to Plato's dualism, in which the forms exist apart from their material expressions, and matter and spirit are capable of independent existence. These differing positions have influenced the history of Christian theology in terms of the way in which both nature and human nature have been perceived. How the natural world is understood, physically and spiritually, requires a clear definition in any move towards a contemporary theology of creation.

Augustine and Aquinas

In the 13th century St Thomas Aquinas, building upon the philosophy of Aristotle, affirmed a unity and interdependence of soul and body and believed that divine grace was a perfecting of nature. Aquinas considered reason to be part of the nature of God and of humanity: 'The fact that we do learn everything we know is a consequence of *our* nature, not of the nature of knowledge' (Kretzmann 1997, p. 1). He was confident about the power of humans to understand the cosmos and the importance of the body is affirmed in his belief that there is nothing in the intellect that was not first in the senses. Since human beings are part of nature, he believed them, like the cosmos, to be rational and the world intelligible: he saw a unity between humanity and the things it sets out to understand. Aquinas believed that, through the intellect, man could recognise the intelligible aspects of sense objects, discover the universal essence of things and, ultimately, contemplate the nature of God. Such beliefs have been instrumental in the success of Western science. An assumption that the cosmos is comprehensible has fired the endeavour to unravel the complexities of nature and discover an underlying system of laws which govern the mechanism of the universe. It may be argued that the desire to reconcile science and religion has its roots firmly embedded in the relationship between humanity and nature as expressed in Thomism.

In contrast, the earlier work of Augustine displays a Platonic dualism. Augustine's emphasis upon the Fall promoted the idea of a sacredness that was separate from the fallen-ness and corruptibility of nature. The place of human reason as a path to an understanding of God's nature was seen as suspect, reason being understood as part of the essentially flawed state of the human condition. Humanity, understood in this way, is in need of transformation by divine grace and, by believing in its essential state of wretchedness, it may be claimed that human nature and all earthly reality is placed in danger of assuming an identity separated from God and consequently divorced from any supernatural mark or purpose. Thus, the positions of Aquinas and Augustine present markedly different perspectives upon the natural order, the former seeing divine grace as a perfecting of an intrinsic goodness present in creation, and the latter seeing divine grace as a necessary means of transformation from without of an essentially flawed and sinful world. It is commonly supposed that science has been instrumental in the development of a purely material understanding of reality but, as will be shown, theology has played its own part, and may be said to have been an unintentional partner in the formation of a view of the natural order from a purely secular perspective.

Nature and divine grace in history

Until the 16th century the common view of nature was that of a creation permeated by the imprint of the divine. In the early centuries of Christianity the Church Fathers distinguished between 'image' and 'likeness' of God, the former affording humankind a dignity through being God's own creation, and the latter being developed as human beings worked towards their supernatural end. Aquinas's somewhat optimistic view of human nature saw human beings as created for this God-given end, achievable through the gift of reason and the help of divine grace. Aquinas believed reason to be part of the nature of both God and humanity. The natural law was thus seen as a product of man's rational nature and the rational creature's participation in the eternal law which underpins God's intentions for man and creation. Human nature, understood in this way, possesses an affinity for relationship with God through its rational understanding of a natural order founded upon the God-ordained laws which determine creation's purpose and eschatological significance. Based upon his notion of the unity of body and soul and the place of reason in man's search for God, Aquinas extended a certain principle from Aristotle to affirm a special place for humanity in relation to God. Aristotle believed there can be no natural desire for anything that cannot be achieved by natural means. However, Aquinas made human nature an exception: man has a natural desire for that which he *cannot* attain – the beatific vision or vision of God.

In the sixteenth century, in a movement which came to be known as 'neo-scholasticism', there was a revival of Aquinas's teaching. Two particular scholars had a powerful influence upon its interpretation, and their works formed the foundation of the nature and grace relationship for the subsequent centuries. Unlike Aquinas, they made no exception for mankind, thus establishing the idea of natural desires with only natural ends (Milbank 2005, pp. 16-17). They were Thomas de Vito, a Dominican cardinal known as Catejan, and Francisco de Suarez, a Jesuit theologian from Spain (McPartlan 1995, p. 47). They have since been charged with distorting the teaching of St Thomas, and their work described as 'a travesty of the teaching of the master....' (McPartlan 1995, p. 48). Human nature, for Catejan and Suarez, possessed no natural longing for the supernatural: human beings may be elevated only by the will of God, should God implant such a desire in an individual soul. This upheld the sovereignty of God, the argument being that, if man does possess a natural desire for God, God is then placed under some obligation to meet that desire in order for human beings to fulfil their nature: God cannot be said to have obligations to man, therefore human beings cannot possess a natural desire for God. Seen this way, divine grace 'does not elevate nature in such a way that it further develops the natural' but, instead, becomes 'a kind of purely nominal change in status by the decree of an arbitrary God' (Milbank 2005, p. 22).

Implications of neo-scholasticism

This new reading of Aquinas on nature and grace came to dominate all subsequent theology: the idea of 'pure nature' devoid of any mark of the supernatural and '*in actuality* ... fully definable in natural terms' (Milbank 2005, p. 17). The realm of super-nature became the concern of theology and that of nature the remit of philosophy: nature and a natural end 'became a closed realm or tier, wherein natural desires are naturally fulfilled' (McPartlan 1995, p. 50). Other developments in cultural history have, of course, played their part. Renaissance humanism and the emergence of more scientific interpretations of the world through the Enlightenment have tended to view creation from a material perspective. The dualism of Augustine and the influence of many Protestant, particularly Calvinist, thinkers on the flawed, or even depraved, state of nature as a result of the Fall have played their part, causing a further separation of that which is holy or sacred from the world into its own protected enclave. But the dualism of nature and the supernatural, instigated by neo-scholastic theology in the sixteenth century, and promoted by the Church, could only have contributed to the exposure of the world to a purely natural interpretation.

De Lubac and the 'new theology'

Just as neo-scholastic thinking was beginning to wane in the 19th century, Pope Leo XIII, in an encyclical letter, *Aeterni Patris* in 1879, ruled that neo-scholasticism was to be the only philosophy and theology taught in Catholic seminaries. He also encouraged biblical exegesis and patristic research without realising their potential to challenge the very theology he was promoting. Through this, Aquinas's thought was reviewed again, including 'his repeated emphasis on the goodness of created nature' (Kerr 2002, p. 5). The French theologian Maurice Blondel has been cited as providing the impulse for a return to theology's more authentic tradition, and a revival of the real Aquinas (McPartlan 1995, p. 51). Following this, the French Jesuit Henri de Lubac, in a ground-breaking publication *Surnaturel* in 1946, sparked off what was described as 'the most bitter controversy within twentieth century Thomism, and in Catholic theology at large' (Kerr 2002, p. 134). In what was termed the 'new theology', a movement prior to Vatican II, there was sought a new relationship between creation and divine grace.

De Lubac's picture of nature was theocentric, recognising everything that human beings call 'natural' as both created and sustained by God. He believed that man, in losing sight of this view, comes to have an anthropocentric perspective on creation in which 'natural' refers to material being and self-sufficiency. De Lubac did not believe that a natural desire for God in any way compromised the transcendent character of grace or the freedom of God to bestow grace upon humanity. He held that, in protecting the transcendence of grace, the neo-scholastics had turned it into something separate from any property intrinsic to human beings. If a desire for union with God is in nature, God's own fulfilment of that desire is a fulfilment of his own summons: thus the desire is God's own gift (Duffy 1992, p. 71). God's will to give Godself means that humanity is created to desire God and that 'through acquiescence to it, nature obeys the basic ontological orientation implanted in it by the Creator' (Duffy 1992, p. 72). Characteristic of de Lubac's unifying approach to nature and the supernatural and his teleological view of humanity, he regarded the Aristotelian and Thomistic unity of body and spirit as a greater protection and affirmation of bodily value and spirituality than the dualism of Platonic and Cartesian philosophy (Milbank 2005, p. 18). The human being, in this sense, is a synthesis of matter and form in which the immaterial form determines both what something is and what it will become.

De Lubac believed that Catejan's reduction of man to a merely natural being had produced a dichotomy between man's natural and supernatural end. In holding that the supernatural defines the natural, he believed it was the call of grace that defined our humanity and that humans were created for a single end, a synthesis of a double gratuity of creation and destiny. He did not believe that man could achieve this for himself but, in his acknowledgement of human beings as destined by nature for a supernatural life, he saw the human response to God as resulting from a supernatural summons by which all human beings are called. Man's supernatural destiny, to de Lubac, cannot be frustrated unless man turns away from God by his own free choice, and this cannot occur without an essential suffering within the being who denies his own nature and destiny. The finality for which human beings are created was, to de Lubac, inscribed upon their being as part of the universe created by God, and he affirms his own nature as one assigned by God for supernatural finality: 'there is only one end, and therefore I bear within me, consciously or otherwise, a "natural desire" for it' (de Lubac 1967, p. 72).

De Lubac's theology of nature and grace is one that continues to provoke controversy. The thought that fallen and sinful humanity might have a natural desire for the vision of God (albeit formed and fulfilled by divine grace alone) has not been well received in certain circles (Duffy 1992, p. 67). However, behind de Lubac's apparent elevation of nature, and what may be seen as an invitation to a more individualistic relationship between the human being and God, there lies a notion of the created order that is not merely life-affirming but God-affirming, in that creation, being imbued with the supernatural, becomes once more seen as the object of God's creative and sustaining action.

Rahner and the 'supernatural existential'

In spite of the suspicions surrounding his work, de Lubac participated in Vatican II in which the new theology helped to create a greater openness to the world and an acknowledgement of the spiritual value of the world outside the Church. The German theologian Karl Rahner developed his own theology of nature and grace, steering a careful path between the new theology and orthodoxy, to produce an acceptable compromise. Rahner, like de Lubac, endeavoured to counter the idea of grace as coming from without, an accidental addition or imposed second level on human nature which, without this addition, was seen as sufficient and complete in itself. Rahner assumed human beings to be created for relationship with God in both the earthly life and eschatological destiny. His notion of the 'supernatural existential', a property of human beings which is open to transcendental reflection, was presented as a synthesis of the call of God and God's provision, within the human being, of the desire for union and the capacity for response. He believed that, if it is God's will that all should be saved and come to a knowledge of God, 'God's grace must be present throughout creation' (Rahner 2004, p. 15). He thus proposed the view that humanity's path to God is through nature, grace being nurtured through man's interaction with the world of experience.

Nature and grace today

The theologies of de Lubac and Rahner form important landmarks in the theological reflection upon the nature of man and his place in the natural world. Both stood against the concept of 'pure nature' and the loss of the idea of man as both a natural *and* a supernatural being which they considered as essential to any description of the human condition. The affirmation of this view may be said to have a particular urgency in a social environment where human beings and the world they inhabit are increasingly understood in purely material terms. God is the God of all concrete reality, and if the grace of God is seen as permeating all the created order it will involve every part, from the inmost longing of the individual to the most secular of human activities. Thus, a greater 'supernaturalising' of the natural world restores creation to that of an order with a natural, a supernatural and a teleological significance: nature is elevated above the concept of the purely natural and protected from a subsequent exposure to the pure materialism of scientific naturalism and secular humanism.

Christological perspectives

Such attempts to bring a supernatural significance to the natural world do not, however, necessarily lead to the God of the Judeo-Christian tradition. The spirituality of the created order has relevance in various religions which may either acknowledge an objective deity or be mere philosophical extensions of the inner 'spirituality' of the human mind. The God of the Holy Trinity defines the God-human relationship in a way that uniquely draws humanity into divine relationship through a God who lived among us. The Gospels give weight to the idea of earthly experience as the starting point of man's discovery of divine mystery, since they mark the encounter of God with the concrete reality of the natural world. De Lubac and Rahner saw the mystery of the Incarnation as central to the relationship of nature and grace, a theology that undervalues neither the divine gift of God nor the created order. There would thus appear to be no reason to assume that a desire for transcendence and a spirit of world-affirmation need necessarily run counter to one another; indeed, both de Lubac and Rahner point to these as existing together in the notions of man's reach for God and the gracing of created life. It may be argued that, as an incarnational religion, Christianity's unique understanding of God's transcendence and immanence provides the foundation for a theology of creation which both protects what is orthodox from a slide towards pantheism and celebrates the created order in a way that speaks to contemporary society. Nature and grace, as fully expressed in the God who took upon himself our human state, reveal both the mystery of grace and the fulfilment of nature.

Bibliography

Catechism of the Catholic Church. 1994. English translation from Latin text copyright, Liberia Editrice Vaticana. London: Geoffrey Chapman.

De Lubac, H. 1967. *The Mystery of the Supernatural,* tr. Rosemary Sheed. London: Herder & Herder.

Duffy, S. 1992. *The Graced Horizon: Nature and Grace in Modern Catholic Thought.* Collegeville, MN: The Liturgical Press.

Kerr, F. 2002. *After Aquinas: Versions of Thomism.* Oxford: Blackwell.

Kretzmann, N. 1997. *The Metaphysics of Theism: Aquinas's Natural Theology in Summa Contra Gentiles.* Oxford: Clarendon Press.

McPartlan, P. 1995. *Sacrament of Salvation: An Introduction to Eucharistic Ecclesiology.* Edinburgh: T&T Clark.

Milbank, J. 2005. *The Suspended Middle.* Grand Rapids, Michigan: Eerdmanns.

Rahner, K. 2004. *Spiritual Writings,* ed. Philip Endean. Maryknoll, NY: Orbis Books.

CHAPTER TWELVE

GOD AND THE MATTER DELUSION: THE DENIAL OF MATTER IN THE TEACHINGS AND PRACTICE OF CHRISTIAN SCIENCE

DANIEL R. D. SCOTT,
INDEPENDENT SCHOLAR AND MEMBER OF THE
CHRISTIAN SCIENCE BOARD OF LECTURESHIP

Introduction

This paper explores the understanding of matter within Christian Science – the metaphysical and theological system set forth in *Science and Health with Key to the Scriptures* by Mary Baker Eddy, and taught by the Christian Science Church she founded.

The Christian Science worldview is of particular interest in the context of this book, because it is diametrically opposed to that of atheist materialism (Eddy 1910a, pp. 349-350). In the radical ontology of Christian Science, God (understood as Spirit) and the spiritual creation are seen as the sole reality of existence, while matter and the material sense of existence are understood ultimately as delusions.

Background

It is beyond the scope of this paper to explore in any detail how Mary Baker Eddy arrived at her radical and unorthodox conclusions about matter and the nature of reality. However, some background information about her life will help to put her conclusions into context.

Mary Baker Eddy was born was born in 1821 in New Hampshire. She was raised by a strict Calvinist father, who taught predestation and eternal damnation for sinners, and by a mother who instilled the idea of God being Love. Living in New England in the nineteenth century, Eddy grew up in the same climate of thought that produced the transcendentalists Ralph Waldo Emerson and Bronson Alcott. In later life Eddy struck up a friendly correspondence with Alcott, who developed a great affinity for Eddy and her views. However, she did not meet or correspond with these thinkers until after she had formed and articulated her own ideas.

While she was relatively well-educated for a woman of her time, it does not seem possible to trace any substantial influence to any previous thinkers, excepting the Biblical authors. (Modern scholarship dismisses the idea that she appropriated the ideas of P. P. Quimby (see Gill 1999), though some books still perpetuate this fallacy.) Indeed, Eddy cites the Bible as her sole authority (Eddy 1910a, p. 126) – though her conclusions were not the result of an abstract theological investigation, but rather the result of a practical search for health and healing. While Eddy acknowledged that her denial of the validity of mortal existence (and its sin, sickness and death) ran completely contrary to the testimony of physical sense, she quoted St Paul in attributing her 'heavenly conviction' (Eddy 1910a, p. 108) to 'the gift of the grace of God given unto me by the effectual working of His power' (cf. Ephesians 3:17, KJV).

Although Mary Baker Eddy's search began and ended with the Bible, along the way she explored homeopathy and other healing practices. These caused her to doubt the efficacy of matter as a healing agent, and finally to conclude that the human mind was more primary than matter. Her 'falling apple' came in February 1866 when she was herself healed (through reading and pondering a New Testament healing) of the results of a fall that was considered life threatening (Eddy 1891, p. 24). At the time she did not know how she had been healed, but she was determined to find out. She later explained: 'I knew the Principle of all harmonious Mind-action to be God, and that cures were produced in primitive Christian healing by holy, uplifting faith; but I must know the Science of this healing, and I won my way to absolute conclusions through divine revelation, reason, and demonstration' (Eddy 1910a, p. 109).

The questions she may have asked herself are still germane to the science and religion dialogue today: How do we make sense of miracles? How did Jesus and his disciples heal? And, What is the Science behind Christian healing?

In 1875 Eddy published the first edition of her primary work *Science and Health*, and she spent the rest of her long life rewriting and revising this book. She saw it as the definitive statement of her teachings, and went on to 'ordain' the Bible and it as the pastor of the church she established (Eddy 1910b, Article 14 section 1). For this reason, any serious examination of the teachings (and practice) of Christian Science involves study of this primary text as well as her other published writings.

A radical ontology

Science and Health sets forth some startling and unusual ideas. Eddy writes:

> My discovery, that erring, mortal, misnamed *mind* produces all the organism and action of the mortal body, set my thoughts to work in new channels, and led up to my demonstration of the proposition that Mind is All and matter is naught as the leading factor in Mind-science (Eddy 1910a, p. 108).

It should be noted that at the end of this passage 'Mind' is capitalised, and used as a term for God. So Eddy is saying that God is All – but not in the usual pantheistic sense, as evidenced by the simultaneous claim that matter is nothing (cf. Eddy 1898).

It is hard to know exactly how Eddy came up with such a radical ontology, but if we are to take her at her word her pivotal healing experience in 1866 gave her a glimpse that 'Life in and of Spirit' is the 'sole reality of existence' (Eddy 1896, p. 24). After this experience, Eddy came to see God's allness and goodness as self-evident propositions (Eddy 1910a, p. 113), and concluded that the universe of God's creating is entirely spiritual and therefore entirely perfect. At the same time she maintained that mortal sense of existence is an illusion – a 'waking dream' (Eddy 1910a, p. 250), and that mortal belief mistakes this fiction for fact. She explains: 'The realities of being, its normal action, and the origin of all things are unseen to mortal sense; whereas the unreal and imitative movements of mortal belief, which would reverse the immortal modus and action, are styled the real' (Eddy 1910a, p. 212).

Throughout her writings Mary Baker Eddy contrasts the spiritual and eternal with the material and mortal. In her view the spiritual and eternal represent reality (the universe as God sees and knows it), while the material and mortal are counterfeits of reality – errors of sense – and are illusions, in as much as they appear real to human consciousness but have no fundamental reality. Mary Baker Eddy summed up her worldview in a short passage she referred to as the *Scientific Statement of Being*. It reads:

> There is no life, truth, intelligence, nor substance in matter. All is infinite Mind and its infinite manifestation, for God is All-in-all. Spirit is immortal Truth; matter is mortal error. Spirit is the real and eternal; matter is the

unreal and temporal. Spirit is God, and man is His image and likeness. Therefore man is not material; he is spiritual (Eddy 1910a, p. 468).

The *Scientific Statement of Being* is found in the chapter 'Recapitulation' in *Science and Health*. This chapter is based on Eddy's original class teaching notes, and is still used today in the church's formal teaching of Christian Science. Thus it is clear that the *Scientific Statement of Being*, with its denial of matter and its affirmation of spiritual reality, is core to the teachings of Christian Science, and key to understanding them. For Eddy, this statement was so important that she requested it be read at the end of every Sunday Service in all Christian Science churches, together with 1 John 3:1-3.

Understanding matter within the Christian Science worldview

Christian Science is perhaps unique among religious movements in the emphasis it places on understanding the nature of matter: the word 'matter' appears 739 times in *Science and Health*. In the *Scientific Statement of Being*, matter is identified with 'mortal error' and 'the unreal and temporal'. Elsewhere Mary Baker Eddy variously describes or defines matter as a 'human concept', a 'fiction' and an 'error of statement', 'Mythology', 'mortality', 'illusion', 'the opposite of Truth', 'the opposite of Spirit', 'the opposite of God' (Eddy 1910a, pp. 277, 469, 170, 591). Possibly most illuminating are Eddy's explanations of matter as 'the subjective state of what is termed by the author mortal mind', and 'that which mortal mind sees, feels, hears, tastes, and smells only in belief' – although these need further explanation.

'Mortal mind' is a term that Eddy introduced to describe the human mind (in contradistinction to the divine Mind), in which matter seems as real as Spirit (God), and evil as real as good. However, she explained that 'as the phrase [mortal mind] is used in teaching Christian Science, it is meant to designate that which has no real existence' (Eddy 1910a, p. 114). Nevertheless, Eddy did not deny human experience, and maintained: 'There is but one Creator and one creation' (Eddy 1910a, p. 502). In Eddy's view, however, the true nature of this one spiritual creation was obscured by the false material sense of things, and must be demonstrated step by step in life practice. Eddy's pragmatism is apparent in the following passage:

During the sensual ages, absolute Christian Science may not be achieved prior to the change called death, for we have not the power to demonstrate what we do not understand. But the human self must be evangelized. This task God demands us to accept lovingly to-day, and to abandon so far as practical the material, and to work out the spiritual which determines the outward and actual (Eddy 1910a, p. 294).

One way Eddy explained the nature of human experience was with reference to Jesus' parable of the tares and the wheat (Matthew 13:25-40). For her, the wheat represents the spiritual, eternal and real, while the tares represent the mortal, temporal and, ultimately, unreal. Human experiences seem to be a mixture of the tares and the wheat – the real and the unreal – but they 'never really mingle' (Eddy 1910a, p. 300), and are inevitably separated at the time of harvest.

Material science

Given the Christian Science understanding of matter as a mortal illusion, it is interesting to consider Eddy's views of material science. Eddy writes: 'What are termed natural science and material laws are the objective states of mortal mind. The physical universe expresses the conscious and unconscious thoughts of mortals. Physical force and mortal mind are one' (Eddy 1910a, p. 484). Here there seem to be at least some similarities between Eddy's understanding of natural science and the much more recent work of Bruno Latour. By drawing parallels between science as it is really practiced and pre-modernist science, Latour has also found a correlation between natural science and basic actions of the human mind (Latour 2005). Other constructivist philosophies of science also suggest that the physical entities of science are not objective discoveries of independent realities, but are instead in some sense constructions of the scientific enterprise itself (Golinski 2005). Within cosmology, the various versions of the anthropic principle proposed by Carter (Carter 1974) and by Barrow and Tipler (Barrow and Tipler 1986) suggest that the very presence of observers in the universe places some constraints on the observed laws and fundamental constants of physics (which appear to be fine-tuned for biological life). In its weaker forms this is simply a selection effect, but stronger versions speculate that the universe is such that we *must* exist; and John Wheeler has speculated that we participate in determining the observations themselves (Wheeler 2006).

A century earlier, Eddy posited what might in retrospect be considered a kind of constructivist anthropic principle, when she wrote: 'Whatever theory may be adopted by general mortal thought to account for human origin, that theory is sure to become the signal for the appearance of its method in finite forms and operations' (Eddy 1910a, p. 553). Hilary Lawson has written of the problem raised by self-reference in both modern and post-modern accounts of experience (e.g. Lawson 1985). Eddy was also aware of the reflexive or self-referential nature of mortal phenomena, writing: 'Mortal mind produces its own phenomena, and then charges them to something else, – like a kitten glancing into the mirror at itself and thinking it sees another kitten' (Eddy 1910a, p. 220).

It is noteworthy here that Eddy saw mortal mind as producing its *own* phenomena. This helps to explain one of the most subtle questions in the metaphysics of Christian Science: Where does error, or the mortal dream, originate? In the teachings of Christian Science error has no cause outside of itself. *Science and Health* explains: 'The history of error is a dream-narrative. The dream has no reality, no intelligence, no mind; therefore the dreamer and dream are one, for neither is true nor real' (Eddy 1910a, p. 530). In this way Christian Science explains away the problem, by asserting that the dream for which we are seeking a cause actually has no real existence. And being neither true, nor real, the dream and the dreamer are in fact nothing. Eddy describes mortal mind (and by extension the mortal sense of existence) as 'nothing claiming to be something' (Eddy 1910a, p. 591). Interestingly there are some similarities here with philosopher Hilary Lawson's post-modern metaphysics of *closure* (cf. Lawson 2001). In his *Closure: A Story of Everything*, Lawson suggests that the fundamental natures of matter and human experience are not eternal realities that scientists discover, but somewhat arbitrary results of the human activity he calls 'closure' arising from a 'no thing' which he calls openness. In both Lawson's and Eddy's work, mortal experience is something of an educated *belief* (Eddy 1910a, p. 194).

Where Eddy departs from Lawson is that in Christian Science these non-realist views apply only to the mortal sense of existence – 'the world of error' (Eddy 1910a, p. 13). In Christian Science 'the world of Truth' is seen as having an eternal and indisputable reality. Eddy was keen to point out the 'frailty and inadequacy' (Eddy 1910a, p. 194) of mortal beliefs in order to turn thought to the immortal and real, namely God and the spiritual creation. Christian Scientists believe that this transformation of thought from a material to a spiritual basis has a wonderful side-effect: healing.

Practice

The Christian Science understanding of the allness of God and the nothingness of matter is a key point in the healing practice of its adherents. If the premise that matter and mortal existence are ultimately illusions is held to be true, as Christian Science teaches, and reality is actually spiritual and perfect, then discord and disease can be challenged and healed on this basis. Christian Scientists consider their practice of Christian healing to be a direct continuation of the healing practiced in the Bible, described by St Paul as 'casting down imaginations' (II Corinthians 10:5).

The last 100 pages of Mary Baker Eddy's primary work *Science and Health with Key to the Scriptures* are filled with testimonies of healing from those who have put into practice Eddy's teachings. Testimonies include healings of lung diseases, alcohol and tobacco addictions, deafness and eyesight problems. For over a hundred years verified testimonies of healing have been published each week and each month in magazines published by the Christian Science Publishing Society.

The Church of Christ, Scientist, tends to shy away from the statistical analysis of reports of healing. Nevertheless, in the past some analysis has been made of content of published testimonies. One study (Phinney 1990) looked at all the testimonies of healing published in the Christian Science Journal and Christian Science Sentinel between 1971 and 1981, and found:

- During that 10 year period there were well over 4,000 testimonies of healing and spiritual regeneration published.
- About 1,400 of these were healings of specifically named physical problems.
- They included healings of such ailments as heart disease, cataracts, broken bones, epilepsy, diabetes, cancer, spinal meningitis, and others.
- 46% of these cases had been medically diagnosed.
- Many of these healings were later confirmed by a follow-up diagnosis.
- Of the children's cases published, 51% had been medically diagnosed.

Robert Peel, whose book *Spiritual Healing in a Scientific Age* presents both a thoughtful discussion of spiritual healing and a number of case-studies, writes:

Christian healing is possibly the point at which the supremacy of matter is most directly challenged. Even in the modified form in which doctor and pastor team up to supplement technology with prayer, two different sorts of universe are implied though not acknowledged. Beautiful results may sometimes follow from the cooperation with a doctor who himself believes deeply in Christian Prayer, but a central ambiguity remains a permanent weakness in such an undertaking. Pragmatically the two approaches may work; logically they represent incompatible theories of cause and effect (Peel 1987, p. 196).

The history of science has furnished us with different models of matter and reality. As scientists and/or people of faith we have to decide which models of reality are valid and meaningful to us. In a paragraph with marginal heading 'Choose ye to-day', Mary Baker Eddy presented a stark choice to her readers.

Dear reader, which mind-picture or externalized thought shall be real to you, – the material or the spiritual? Both you cannot have. You are bringing out your own ideal. This ideal is either temporal or eternal. Either Spirit or matter is your model. If you try to have two models, then you practically have none. Like a pendulum in a clock, you will be thrown back and forth, striking the ribs of matter and swinging between the real and the unreal (Eddy 1910a, p. 360).

This warning against trying to work from two contradictory standpoints at the same time helps explain why adherents see Christian Science as alternative, rather than complimentary medicine, and hence why they generally choose to seek *either* a medical solution *or* to rely wholeheartedly on Christian Science, but tend to avoid using both approaches at the same time. This resonates with Kuhn's insistence that two different paradigms cannot be adhered to simultaneously (cf. Kuhn 1968).

Conclusion

The idea that matter is a delusion may still seem startling to some. However, if one takes the ideas of God and heaven seriously, it is surely axiomatic that in heaven substance must be spiritual and eternal, and not material and temporal. Hence, the transitory, mutable and temporal nature of matter suggests that matter is a heavenly impossibility, for that which is temporal and mutable cannot be eternal and immutable. So, from a heavenly perspective, it follows that matter is an illusion or delusion. Christian Science asserts that, despite the evidence of the physical senses, this heavenly model of reality is ultimately the only valid model of reality.

In looking beyond the visible universe of matter models to the deeper reality of the invisible universe and its eternal spiritual ideals, Christian Scientists join with Paul in affirming that '...we look not at the things which are seen, but at the things which are not seen: for the things which are seen are temporal; but the things which are not seen are eternal' (2 Corinthians 4:18, KJV).

Bibliography

Barrow, J. D. and Tipler, F. J. 1986. *The Anthropic Cosmological Principle.* Oxford: Oxford University Press.

Carter, B. 1974. 'Large Number Coincidences and the Anthropic Principle in Cosmology' in *IAU Sympsium 63: Confrontation of Cosmological Theories with Observational Data*, pp. 291-298. Dordrecht: Reidel.

Eddy, Mary Baker 1891. *Retrospection and Introspection.* Boston: Christian Science Board of Directors.

—. 1896. *Miscellaneous Writings 1883-1896.* Boston: Christian Science Board of Directors.

—. 1898. *Christian Science versus Pantheism.* Boston: Christian Science Board of Directors.

—. 1910a. *Science and Health with Key to the Scriptures.* Boston: Christian Science Board of Directors.

—. 1910b. *Manual of the Mother Church The First Church of Christ, Scientist, in Boston, Massachusetts.* Boston: Christian Science Board of Directors.

Gill, G. 1999. *Mary Baker Eddy* (Part of Radcliffe Biography Series). Reading, Massachusetts: Perseus Books / Da Capo.

Golinski, J. 2005. *Making Natural Knowledge: Constructivism and the History of Science.* Chicago: University of Chicago Press.

Kuhn, T. 1968. *The Structure of Scientific Revolutions.* Chicago: University of Chicago Press.

Kukla, A. 2000. *Social Constructivism and the Philosophy of Science.* London: Routledge.

Latour, B. 2005. *Reassembling the Social: An Introduction to Actor Network Theory.* Oxford: Blackwell.

Lawson, H. 1985. *Reflexivity: The post-modern predicament.* London: HarperCollins.

Lawson, H. 2001. *Closure: A Story of Everything.* London: Routledge.

Peel, R. 1987. *Spiritual Healing in a Scientific Age.* London: Harper and Row.

Phinney, A. W. 1990. 'The Spirituality of Mankind' in *Christian Science: A sourcebook of contemporary materials*. Boston: Christian Science Publishing Society.

Wheeler, J. A. 2006. Science Show: The Anthropic Universe. Australian Broadcasting Company, 18th February 2006. www.abc.net.au/rn/scienceshow/stories/2006/1572643.htm (accessed 29 December 2009).

CHAPTER THIRTEEN

ON THE CAPACITY OF SOUND WAVES, PAINTED CANVAS AND PRINTED PAGE TO CARRY MEANING – AND THE PLACE OF THE ARTS IN A NEW-STYLE NATURAL THEOLOGY[1]

PETER BARRETT, UNIVERSITY OF KWAZULU-NATAL, DURBAN

Introduction

The case for theology's engagement with the arts – a recurring task – was argued compellingly by Howard Root (1962, pp. 3-19) in the early 1960s. For Root the main challenge faced by theologians was 'to regain contact with the mind and imagination of the most sensitive segment of our society, that of the poet or novelist or dramatist or film producer.' Perhaps the foremost current study of the interplay between theology and the arts, at least in the West, is that which is conducted by the Institute for Theology, Imagination and the Arts at the University of St Andrews, Scotland, set up by Jeremy Begbie and Trevor Hart in the year 2000.

The challenge articulated by Root is also addressed in the late Anthony Monti's *A Natural Theology of the Arts* (2003). But here his theological approach to the arts developed as an extension of contemporary science-and-theology discourse. The book is based on and inspired by the writings of John Polkinghorne, which Monti studied perceptively and appreciatively. With a strong grounding in English literature and a substantial knowledge

---

of theology's links with philosophy and aesthetics, Monti saw these writings as a resource to be used against the prevailing relativism in the deconstructionist world of literary studies – and, indeed, as a launching pad for the writing of his book.

He was also drawn to George Steiner's *Real Presences* (1989), with its thesis that it is the 'real presence' of God that underlies great works of art: a book filled with impressive insights, yet criticized by Robert Carroll (1994, p. 266), for example, for being short of 'argument rather than assertion, warrants rather than wagers'. It needed a systematic conceptual framework, and Monti responded by developing a natural theology that is carefully argued throughout its epistemological, metaphysical, aesthetic and theological phases.

In this task he builds upon Polkinghorne's distinctive views on the meaning of critical realism and the question whether macroscopic physical systems are characterized by indeterminacy (an ontological property here referred to as 'macro-indeterminacy') or merely unpredictability (an epistemic limitation). In the next two sections we consider these and other key ideas used by Monti, including Michael Polanyi's concept of 'tacit knowledge', before discussing the emergence of a widely embracing new-style natural theology that includes a theology of the arts as one of its components.

Critical realism, macro-indeterminacy and mind/ matter monism

Polkinghorne's views on these three topics stem from his experience of quantum physics over its three lively decades of 'bafflement and break-through', 1950-1980. (This stimulating era of particle physics is described in *Rochester Roundabout* (Polkinghorne 1989), which provides a detailed epistemological discussion in its closing chapter.) In this enterprise physicists became convinced that they had achieved a high degree of verisimilitude between theory and physical reality. The conviction that science unveils something real is common to almost all scientists – their approach is that of scientific realism. But insofar as acts of personal judgement are involved in scientific research it is also a critical realism. N. T. Wright (1992, pp. 35-36) describes this as 'an approach that acknowledges that the only access we have to reality lies along the spiralling path of appropriate dialogue between the knower and that which is known' – which includes, of course, a recognition of the subjective aspect of the process, and of the provisionality of the knowledge gained.

However, with the experience of quantum physics as his starting point, Polkinghorne goes beyond this minimalist description of critical realism, endorsing the notion that in the search to understand any part of reality it is neither sensibility nor picturability that matters, but intelligibility. He claims that it is not unreasonable to assume that what we find intelligible is a reliable guide to what is the case. The idea is expressed pithily in the phrase he coined and has often been challenged to explain: 'epistemology models ontology' (Polkinghorne 1998, p. 65). He argues, first, that epistemology models ontology even if it does not entail ontology. Then, adopting this metaphysically based aphorism as an epistemological principle, he claims that if it is applied to the quantum realm of elementary particles (as mentioned below), it is equally appropriate to apply it to the classical realm of macroscopic systems.

Polkinghorne describes this critical-realist approach as a strategy of seeking the closest possible alignment between what we know (epistemology) and the way things are (ontology). Quantum physicists did so intuitively when they took Heisenberg's Uncertainty Principle a step further. What was simply a case of unpredictability came to be regarded as one of indeterminacy. This was a metaphysical choice, made in terms of the aesthetic criterion of 'naturalness of explanation'. An alternative, logically impeccable version of quantum theory, David Bohm's deterministic account, was found wanting when weighed against this criterion. It seemed unduly contrived.

With the same criterion of naturalness in mind, Polkinghorne invokes the notion of inherent indeterminacy in macroscopic systems – macro-indeterminacy – as a 'causal joint' where, within the grain of nature's processes, human intentionality and, analogously, divine intentionality may be considered to operate. He does so in the interests of formulating a metaphysical scheme that embraces the range of existence from the physical to the personal. Thus, macro-indeterminacy is a pivotal element in Polkinghorne's conception of divine action in the world, whether in the processes of nature or in the inspiring of human minds.

It is worth noting that macro-indeterminacy is considered implausible by an impressive array of science-and-religion writers: R. Russell, A. Peacocke, N. Murphy, G. Ellis, E. McMullin, P. Clayton, T. Tracy and others. Most of them have proposed quantum indeterminacy instead, as the way in for providential divine action, but both Polkinghorne (1998, pp. 59-60) and Peacocke (2001, p. 181) have stood against this, not least because it is not clear how quantum effects may be amplified to take effect at the macroscopic level. Polkinghorne adds that he does not claim that macro-indeterminacy is necessarily the only route for intentionality (the quantum level may also be involved), but he regards the macro-perspective as the more useful and significant (Polkinghorne, private communication); and he stresses (1996, p. 39) that the ultra-sensitivity of any macro-system to small disturbances means that it must be treated holistically, not through the manipulating of micro-events within the system.

The suggestion of macro-indeterminacy stems from the fact that nonlinear macroscopic systems, with their exquisite sensitivity to the slightest perturbation, are not accurately describable by the deterministic equations of chaos theory, unless they are totally isolated from their environment. And since no such system is ever perfectly isolated, the equations cannot describe its behaviour exactly, even in principle – which raises the question put by Polkinghorne (1991, p. 41): Which is to be assigned ontological priority, the behaviour or the equations? Which is the reality? If priority is assigned to the behaviour (*contra* the view of most physicists, suggests Monti (2003, p. 40)), it follows that nature's systems might possess a subtle ontological flexibility which, by definition, the equations cannot show.

An associated metaphysical element in Polkinghorne's approach is the concept of mind and matter comprising a dual-aspect mind/ matter monism rather than a Cartesian dualism – comparable to the dual-aspect wave/ particle monism of quantum physics. At this point his intuition as a physicist provides a fruitful analogy. If a wave-like state is one that contains an indefinite number of particles, then, analogously, we may think of a mental state as associated with an indefinite degree of organization of the brain (not, of course, an indefinite amount of brain matter) – that is, with some flexibility of pattern. He points to the well attested creative activity of the unconscious mind during sleep (Polkinghorne 1988, p. 75), which no doubt stems from its increased flexibility and freedom of association, compared to the more rigidly organized processes of the conscious ego.

Once the twin proposals of macro-indeterminacy and mind/matter monism are accepted, the way is open to the idea of top-down, mind-to-brain input of information – effected holistically and without energy transfer. Polkinghorne (1998, pp. 62-63) suggests that, in some way which may perhaps remain beyond the reach of human understanding, the system is changed by such energy-less 'active' information from one configuration to another, each lying within the scope of its strange attractor – that is, within its set of allowed equal-energy states. He adds that this is no more than a glimmer of how human intentionality might take effect. It could also be how divine intentionality works vis-à-vis the human brain or any other macroscopic system.

For Monti, Polkinghorne's combination of macro-indeterminacy and mind/matter monism, with its consequence of flexible openness of the human mind/brain, is what is needed to begin to probe the concept of artistic creativity at its epistemic roots: to ask how it is that an embodied mind can encounter the mystery of music, art and literature, the realm of metaphor and symbol, and therein find meaning. Such mental activity brings into play what scientist-philosopher Michael Polanyi called 'tacit knowledge', the background to all acts of personal knowing, and the cognitive source at the heart of Monti's scheme. Polanyi's epistemological ideas are discussed in detail in Thomas Torrance's *Transformation and Convergence in the Frame of Knowledge* (1984), from which Monti frequently quotes.

Tacit knowledge, metaphor and multi-levelled reality

The concept of tacit knowledge is Polanyi's main insight into the thought processes involved in scientific discovery and in the build-up of personal knowledge in general, through the exercise of not only reason but also emotion, intuition and imagination. Polanyi's account of the process of scientific discovery is summarized thus by Joan Crewdson:

> The dynamism in scientific discovery is due in part to the deliberate effort or thrust of the imagination searching for clues, in part to the spontaneous activity of intuition which integrates what imagination has hit upon, and in part to the inexhaustible nature of the hidden reality. Polanyi's analysis shows imagination serving the questing mind much as the blind man's stick probes the environment, searching for new coherences and guided by intuition towards a possible solution (Crewdson 1994, p. 58).

Tacit knowledge is the unconsciously held store of knowledge and understanding that develops through the 'conviviality' – the thinking, discussing and working together – within any corporate human activity. And the perceiving of aesthetic qualities intuitively, via tacit knowledge, is what Monti describes as 'the hinge on which critical realism and, indeed, the entire argument of this book turns.' Yet, as Polkinghorne (1989, p. 175) points out, the concept of tacit knowledge has been strangely neglected by philosophers of science – and, we may add, remarkably underplayed in science-and-theology discussions.

Tacit knowledge acts as the 'spectacles behind the eyes' through which the knower sees that which is to be known and apprehended. Often apprehension occurs not by deductive or inductive reasoning, but by a leap of an imagination that has been sympathetically attuned to the subject matter: a leap from the random observation of phenomena to the hypothesis of a pattern (Wright 1992, p. 37), or to artistic insight in the realm of metaphor and symbol.

Wright (1992, p. 40) gives a vivid description of the conveying of something new by means of a metaphor. Such unveiling consists in 'bringing two sets of ideas close together, close enough for a spark to jump, but not too close, so that the spark, in jumping, illuminates for a moment the whole area around, changing perceptions as it does so.'

Monti (2003, p. 61-2) points out that there is another level of integration to be noted: that of the metaphor as a whole and the attuned imagination. Here the imagination indwells the metaphor, thereby gaining access to its meaning. But the meaning, he adds (from Polanyi), reaches out beyond us in an indeterminate range of intelligibility. And neither the jump within the metaphor nor the leap to a hypothesis results from deductive reasoning. Rather, both are imaginative acts of tacit knowing.

The concept of metaphor has been widely discussed and defined (Monti 2003, p. 58). We may think of it first as a literary figure of speech but, in a broader sense, it may also take the form of a painting, or a piece of music, each with its interweaving of elements or images: that is, we may think of a work of art as metaphoric in that it conveys meaning through what it is as a whole, through the imaginatively observed interplay of its elements. Polkinghorne (1994, p. 38) remarks that understanding in art and literature comes from the power of the whole, through intuitive grasp rather than detailed argument; and intuitive grasp, he adds, requires the exercise of imagination informed by tacit knowledge. Elsewhere (1991, p. 30) he refers to the capacity of metaphor, like that of art itself, to carry us into realms of thought which would otherwise be inaccessible.

Here Monti (2003, p. 22) invokes Polanyi's concept of a transcendent, multi-levelled reality that beckons us on to unending questioning through its intimations of hidden dimensions of order and meaning.[2] Polanyi claims that as we move through these levels from the more tangible to the more intangible we penetrate to things that seem increasingly real and full

---

[2] George Ellis (2002), for example, postulates the following set of levels of existence: (i) the physical world of energy and matter; (ii) the contents of human consciousness; (iii) the set of possible physical and biological forms (those in our world constituting a subset); (iv) a world of abstract realities such as mathematical forms, physical laws, powerful symbols, etc; (v) the set of values, meanings and purposes underlying the created order; and (vi) the fundamental meta-level of the being of God. Levels (ii) to (vi) are regarded as real in that each can affect what happens at level (i). Works of art are based on (i) and (ii) but are also associated with (v).

of meaning. Then the deepest reality is possessed by the things that are least tangible, and it is at these deeper levels that much artistic creativity is located, especially in the realm of values, meanings and purposes underlying the created order. Furthermore, these non-material realms are commonly understood and described in terms of metaphoric language and imagery which, in turn, are seen through the spectacles of tacit understanding.

Monti thus draws together ideas of critical realism, macro-indeterminacy, mind/ matter monism, tacit knowledge, metaphor and the broad sweep of levels of reality, emphasizing the vital role of metaphor in his scheme. And having considered particular works of art in terms of these ideas, he ends the book with theological discussion in which he refers to works of art as, in the words of W. H. Vanstone, representing 'the creativity of recognition ... directly and explicitly responsive to the creativity of God' (Monti 2003, p. 136-138).

Monti explains (2003, p. 140) that Steiner, with his undifferentiated monistic concept of God, does not entertain any idea of ultimate purpose in the created order and, indeed, thinks of art as essentially the response of counter-creation, that is, as rival to divine creation. Monti on the other hand sees art as co-creation, as contributing to the eschatological goal of 'new creation.'

At this point he connects Polkinghorne's bottom-up metaphysics of flexible openness to the realm of trinitarian eschatology by introducing the top-down concepts of perichoresis, relationality and particularity: 'open transcendentals' that Colin Gunton (1993, p. 142) discusses as aspects of the being of God. Gunton sees such open transcendentals as enabling 'a continuing and in principle unfinished exploration of the universal marks of being ... (whose) value will be found not primarily in their clarity and certainty, but in their suggestiveness and potentiality for being deepened and enriched.' For Monti, these three open transcendentals are exemplified supremely in metaphor and great works of art. In such terms he seeks to make the case for art as co-creation, and thence as affirmation of the 'real presence' of God and anticipation of the new creation.

Altogether, Monti has creatively extrapolated Polkinghorne's epistemological and metaphysical ideas into the realms of aesthetics and theology. His natural theology of the arts illustrates the fruitfulness of the concept of macro-indeterminacy in the search to understand how the astonishing power of human cognition operates at the highest levels of imaginative engagement: with works of art in particular, and with the unfolding of creation itself as an immense work of art. And to the metaphysical idea of the indeterminacy and flexibility of the human mind/ brain we may add the theological notion of artistic inspiration as the light touch of 'active' information given by the Spirit of God. In similar vein, John Taylor (1972, p. 19) speaks of the Spirit as enabling human beings 'not by making us supernaturally strong, but by opening our eyes'.

A new-style natural theology that includes the arts

Natural theology, as traditionally understood, aimed at finding arguments for the existence of God on grounds external to revelation, whether deductively from logic or inductively from the appearance of design in nature's life-forms. It fell into disrepute after the successive blows of the criticisms of Hume and Kant, the explanatory power of Darwin's theory of evolution, and the authoritative and initially uncompromising opposition of Karl Barth. However, a new-style natural theology (NNT) is emerging from contemporary science-and-theology discourse with the aim of producing a cogent theistic account of our multi-levelled evolutionary world. This is an endeavour that stands alongside the work of philosophers of religion on the further development of arguments for the existence of God: alongside, for example, Richard Swinburne's aims to show that the existence of God is more probable than the non-existence of God by estimating a probability rating for each argument and pointing to the consequent cumulative rating (cf. Swinburne 2004). (Note, too, the volume of essays edited by Craig and Moreland (2009).)

In Polkinghorne's view (1991, p. 75) the main task of this revived and revised natural theology, as a metaphysical scheme arising from Christian theology, is to take account of the insights of science, aesthetics, ethics and theology itself, and integrate them into a single, consistent and coherent account of reality. Similarly, for Alister McGrath (2008, p. 3) NNT involves seeing the world described by the sciences through the spectacles of Christian insight. Its task is then to form a widely integrative theological framework of understanding: one that gives a rich account of the nature and meaning of human personhood and the quest for the often hidden expressions of truth, goodness, and beauty.

Keith Ward places natural theology in a broad theistic setting that is not linked specifically to Christian tradition (Ward 2003). But, like Polkinghorne and McGrath, he envisages a wide-ranging engagement with the world, describing it as the attempt to show how science, history, morality and the arts are so related that a total integrating vision of the place of humanity in the universe may be formulated: an attempt, he suggests, that will be more of an imaginative art than an inferential or deductive science.

Thus, NNT seeks to produce a meta-narrative of the universe that addresses questions about the meaning and purpose of it all. In the words of Christopher Mooney:

> the universe that science studies is not a mere sequence but a story, a struggle upwards through matter, life, thought, history, and culture. Only a narrative can really capture what is going on. And it is precisely this need of humans for meaningful narrative that allows theology to complement the causality of science (Mooney 1991, p. 319).

But if such a narrative is to do justice to the full range of human personhood, it will need to link Christian belief about the creation not only to the natural and human sciences but also to the humanities: not only to the realm of logic and reason, but also to the realm of aesthetics and the imagination. Indeed, NNT may be thought of as a two-fold bridging activity: internally creating a narrative framework in terms of theology's reflections on the religious, scientific, aesthetic and moral dimensions of human thought and understanding, and externally linking this framework to the general discourse of academy and society.

Jürgen Moltmann (1985, pp. 58-59) refers to these two aspects of natural theology as an internal hermeneutical function of intellectually underpinning the Christian faith, and an external educative function of relating key theological ideas to more general discourse. But he adds a third: an eschatological function of anticipating 'the knowledge of God in glory' – in other words, providing metaphorical insight into the nature of this world as 'parable of the world to come'. This eschatological dimension is a fundamental aspect of trinitarian theology, even if discounted by those who hold to a philosophically based theism. It leads naturally to the key NNT idea, clearly speculative but profoundly explanatory, that this is a world in the making, one that constitutes the raw material for a new creation yet to be realized.

The table below shows a way of representing the role of NNT in Western thought. Whereas systematic theology looks for coherence within the broad scope of theology itself, NNT aims, in this scheme, not only to provide an internally unifying perspective but also to enhance the engagement between theology and the intellectual world at large. The dotted connecting lines indicate that there is considerable interplay among these elements.

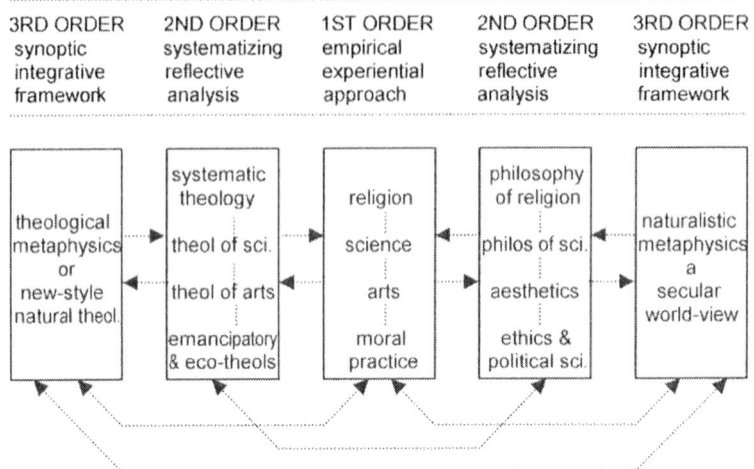

3RD ORDER synoptic integrative framework	2ND ORDER systematizing reflective analysis	1ST ORDER empirical experiential approach	2ND ORDER systematizing reflective analysis	3RD ORDER synoptic integrative framework
theological metaphysics or new-style natural theol.	systematic theology · theol of sci. · theol of arts · emancipatory & eco-theols	religion · science · arts · moral practice	philosophy of religion · philos of sci. · aesthetics · ethics & political sci.	naturalistic metaphysics a secular world-view

Table 1. A rough layout of the structure of Western thought, drawn from discussion with Niels Henrik Gregersen, John Polkinghorne and Keith Ward. The elements are not sharply differentiated; they overlap, and there is much interplay. The two types of second order thinking partly reinforce, partly complement and partly contradict one another, and the two outermost boxes correspond to the extremes of a range of integrative approaches that includes theistic naturalism and process thought.

Such an overarching framework of thought can be formed on the basis of an axiomatic statement about the great drama of creation: on an organizing idea, drawn from either the experientially based, richly detailed theism of one of the monotheistic religions on the one hand, or a more restricted and abstract theism characteristic of philosophy on the other. It seems that within the wide range of science-and-theology discussions, a philosophically based theism is the more frequently assumed or preferred. On the other hand, many find greater explanatory power in the central ideas of Christian tradition, especially in dealing with questions about horrendous suffering (c.f. Adams 1999: Hick 1966) and ultimate meaning.

Then, with early Christian thought about the creation in mind,[3] together with Monti's case for art as co-creation, one can begin to formulate a natural theology on the basis of an organizing idea such as the following:

---

[3] Distinct streams of creation theology have developed from the thought of Irenaeus (c.130–c.203) and Augustine (354-430), respectively. See, for example, John Hick (1966, pp. 207-224) and LeRon Shults (2008, pp. 39-43). Irenaeus

God creates the world with utmost love in order to produce persons who, in communion with God, can freely strive to create the-good-and-the-beautiful (*to kalon*).

But whatever the starting point, *to kalon* and the often invoked divine quality of *kenosis* are key ideas for any NNT in its search to understand something of the immensity of the gracious *creatio continua* underlying this amazing, beloved and costly universe.

In conclusion, NNT is faced with the creative task of drawing together insights from all four areas of reflective theological analysis, not just science-and-theology discussions. The interplay between theology and science over recent decades has been vigorous, and remains important. But the input from the other three areas of engagement, concerning religion, the arts and ethics, will surely make for a well-grounded natural theology: a theological world-view to set against the varieties of fundamentalist thought, secular and religious, and to leaven discussion with the world at large, whether dealing with issues in public life or the big questions of human existence.

Bibliography

Adams, M. 1999. *Horrendous Evils and the Goodness of God*. Ithaca: Cornell University Press.

Carroll, R. 1994. 'On Steiner the theologian', in Nathan A. Scott and Ronald A. Sharp (eds), *Reading George Steiner*. Baltimore MD: Johns Hopkins University Press.

Craig, W. L. and Moreland, J. P. (eds.) 2009. *The Blackwell Companion to Natural Theology*. Oxford: Blackwell.

Crewdson, J. 1994. *Christian Doctrine in the Light of Michael Polanyi's Theory of Personal Knowledge*. Lampeter (UK): Edwin Mellen Press.

Ellis, G. 1993. 'The Theology of the Anthropic Principle', in R. Russell, N. Murphy and C. Isham (eds), *Quantum Cosmology & the Laws of Nature*. Berkely: CTNS/ Vatican Observatory Publications.

—. 2002. 'Natures of existence (temporal and eternal)', in G. Ellis (ed), *The Far Future Universe*. Philadelphia: Templeton Foundation Press.

Gunton, C. 1993. *The One, the Three and the Many*. Cambridge: Cambridge University Press.

---

viewed the creation as good, as fashioned by the divine Word (Christ) and divine Wisdom (the Spirit), with the Spirit acting as Beautifier and Perfecter (Sherry, 2002).

Hick, J. 1966. *Evil and the God of Love*. London: Macmillan.

McGrath, A. 2008. *The Open Secret: A New Vision for Natural Theology*. Oxford: Blackwell.

Moltmann, J. 1985. *God in Creation*. London: SCM.

Monti, A. 2003. *A Natural Theology of the Arts*. Aldershot (UK): Ashgate.

Mooney, C. 1991. 'Theology and Science: A new commitment to dialogue', *Theological Studies* 52. Washington: Georgetown University.

Peacocke, A. 2001. *Paths from Science towards God*. Oxford: Oneworld.

Polkinghorne, J. 1988. *Science and Creation*. London: SPCK.

—. 1989. *Rochester Roundabout*. Harlow UK: Longman.

—. 1991. *Reason and Reality*. London: SPCK.

—. 1994. *Science and Christian Belief*. London: SPCK.

—. 1996. *Scientists as Theologians*. London: SPCK.

—. 1998. *Belief in God in an Age of Science*. New Haven: Yale University Press.

Root, H. 1962. 'Beginning all over again', in A. Vidler (ed), *Soundings: Essays concerning Christian Understanding*. Cambridge: Cambridge University Press.

Sherry, P. 2002 (2nd ed). *Spirit and Beauty: An Introduction to Theological Aesthetics*. London: SCM.

Shults, L. 2008. *Christology and Science*. Aldershot: Ashgate.

Steiner, G. 1989. *Real Presences*. London: Faber & Faber.

Swinburne, R. 2004 (2nd ed). *The Existence of God*. Oxford: Oxford University Press.

Taylor, J. 1972. *The Go-Between God*. London: SCM.

Torrance, T. 1984. *Transformation and Convergence in the Frame of Knowledge*. Belfast: Christian Journals.

Ward, K. 2003. 'Natural Theology', in W. Van Huyssteen (ed), *Encyclopedia of Science and Religion*, pp. 602-605. New York: Macmillan.

Wright, N. T. 1992. *The New Testament and the People of God*. London: SPCK.

CONTRIBUTORS

M. B. Altaie took his Ph.D. from Manchester University. He is now professor of theoretical physics at the University of Yarmouk, Jordan. He specialises in quantum cosmology, and has published many papers on issues in this subject. He is also the author of several books in Arabic on relativity theory, astronomy, and science and religion. His main interest outside physics focuses on the methodology and arguments of Islamic Kalām in the contemporary science and religion debate.

Peter Barrett (PhD Imperial College, London) is a retired Associate Professor who worked in plasma physics for 30 years (10 in the UK and the USA, and 20 in South Africa). His post-retirement years have been devoted largely to science and theology. He is a member of the Anglican Church, and is interested in issues of contemporary Church mission, interfaith engagement, and the possibilities of Christian theology in nation-building. He has published several papers in South African theological journals, written an introductory book on Science and Theology since Copernicus (Continuum 2004) and has helped to organise the South African Science and Religion Forum since its inception in 1993.

Michael Fuller took his Doctorate in organic chemistry at the University of Oxford. He then studied theology at the University of Cambridge and was ordained in the Anglican Church. He is currently Pantonian Professor at the Theological Institute of the Scottish Episcopal Church, and an Honorary Fellow at the University of Edinburgh. He is the author of a book, *Atoms and Icons* (Mowbray 1995), and of more than 20 papers on the contemporary dialogue of science and theology, and on theology and music.

Niels Henrik Gregersen (PhD Copenhagen University, 1987) is professor of systematic theology at Copenhagen University, and co-Director of the Centre of Naturalism and Christian Semantics. His areas of research are contemporary theology and science and religion, with a special emphasis on the science of complexity and on current developments in evolutionary biology. He is the author of four books, and has edited or co-edited fifteen books in science and religion, including

From Complexity to Life (Oxford University Press 2003), *The Gift of Grace* (Fortress Press 2005), and *Information and the Nature of Reality* (Cambridge University Press 2010).

Ruth Gregory was awarded her PhD in Theoretical Physics from the Department of Applied Mathematics and Theoretical Physics in Cambridge in 1988. She then held research positions at the Fermi National Laboratory in Chicago, and the Enrico Fermi Institute at the University of Chicago. She returned to Cambridge in 1993 and in 1995 moved to the University of Durham on a Royal Society University Research Fellowship. After the fellowship she was appointed Reader in Mathematics and Physics in 2003, and Professor in 2005. She was awarded the Institute of Physics Maxwell medal in 2006 for her contributions to Theoretical Physics.

Peter Harrison is Andreas Idreos Professor of Science and Religion at the University of Oxford. A fellow of Harris Manchester College, he is also Director of the Ian Ramsey Centre at Oxford. He has published extensively in the area of cultural and intellectual history, with a focus on the philosophical, scientific and religious thought of the early modern period. His most recent books are *The Fall of Man and the Foundations of Science* (Cambridge 2007) and *The Cambridge Companion to Science and Religion* (Cambridge 2010).

John Henry is a Reader in the history of science in the Science Studies Unit of the University of Edinbuurgh. He has published widely in the history of science from the Middle Ages to the Nineteenth Century, and has a particular interest in the historical relations between science and religion. He has recently published (with John M. Forrester) *Jean Fernel's On the Hidden Causes of Things: Forms, Souls, and Occult Diseases in Renaissance Medicine* (Leiden: E. J. Brill 2005); and the third (revised) edition of his *The Scientific Revolution and the Origins of Modern Science* (London and New York: Palgrave 2008).

James Jones became Bishop of Liverpool in 1998, having been Bishop of Hull since 1994. Over the last fourteen years he has been deeply involved in Urban Regeneration. For four years he chaired the New Deal for Communities programme in Liverpool (Kensington Regeneration) and has championed community-led regeneration in lectures, articles and broadcasts. He broadcasts regularly, especially on 'Thought for the Day' for the BBC. He has written a number of books including 'Jesus and the

Earth' (SPCK 2003) which looks at the relationship between Christianity and the environment. He is a member of the House of Lords, Bishop for Prisons, Visitor to St Peter's College in the University of Oxford, Co-President of Liverpool Hope University and Adjunct Professor teaching an MA on theology and the environment, a Vice President of Tear Fund, WWF Ambassador and a Fellow of the RSA (Royal Society for the encouragement of Arts, Manufactures and Commerce).

Hilary Martin began her career in medical physics at the National Hospital, Queen Square, London, in the 1960s, working in research and development of diagnostic technology for disorders of balance. She later entered the teaching profession, retiring from St Mary's School Ascot in 2003 where she taught physics and also ran the sixth form philosophy society. She completed a Certificate in Theological Studies at the Royal Holloway College in 2004 and an MA in Philosophy and Theology at Heythrop College in 2007.

Michael Poole was Head of the Physics Department at a London Boys' Comprehensive School. After undertaking some broadcasting work on Science and Religion he was appointed Lecturer in Science Education at King's College, University of London. He is currently Visiting Research Fellow in Science & Religion at King's College and is the author of some eighty articles and several books on the subject.

Colin Russell spent nearly twenty years in organic chemistry, and has since specialised in the history of science (especially chemistry) and the relations between science and religion, particularly from an environmental viewpoint. His latest book is *Saving Panet Earth: A Christian Response* (Authentic Media 2008).

Daniel Scott planned to be a theoretical physicist, but after seven years of studying mathematics and mathematical physics at Cambridge University he felt called to the healing ministry of his church, and has spent the last ten years working as a practitioner of Christian Science healing. In 2008 Daniel was appointed to the Christian Science Board of Lectureship, and now travels the world giving lectures on science and religion from a Christian Science perspective. He recently helped defeat the motion 'Science has made religion obsolete' at Durham Union Society.

Kenneth Wilson is a Senior Visiting Research Fellow at the universities of Canterbury Christ Church and Chichester. He worked with Arthur

Peacocke to establish the Ramsey Centre and was a member of the Christ and the Cosmos network. He was Principal of Westminster College, Oxford from 1981-1996 and subsequently established the Research Centre at The Queen's Foundation, Birmingham. He is the Deputy Director of 'Learning for Life', a major research programme into the values, attitudes and dispositions of students covering the age range 3-25. He has written several books in the areas of philosophy, theology and education. He is an ordained Methodist minister.

INDEX

aesthetics 10, 19, 142, 147, 148, 149
alchemy 43-44
al-Ghazali 44, 89
al-Nazzam 91
Anaximander 39
Anaximenes 39
Anderson, C. 16
Anselm, St 113
anthropic principle 135
antimatter 9, 16-17, 19
Aquinas, St Thomas 43, 124ff
   'five ways' of, 50
Aristotle 40-41, 42, 92, 124
atheism 45-46, 63
atomism 10, 39-40, 44, 46, 53, 68ff,
   83
   Arabic 43, 44, 46, 88, 89-90
Augustine, St 41-43, 125, 126,
   150n.

Bacon, F. 44
Barbour, I. 84
Barrow, I. 47-48
Barth, K. 148
Bell's theorem 29
Bentley, R. 47
Bernoulli, D. 60
Bible 1, 80, 113
Big Bang 18-19
Bohm, D. 29, 143
Bohr, N. 9, 12-13, 26, 29, 75, 84, 91
Boltzmann, L. 10
Bonaventure 50
Born, M. 22-23, 26, 29
Boyle, R. 47, 62, 68, 83
Bruno, G. 44
Butlerov, A. M. 73

Cambridge Platonists 46, 48, 62-63

Cappadocian Fathers 111
causal joint 93, 143
chaos theory 144
Clarke, S. 47
Clausius, R. 105
Clifford, W. K. 61
climate change, see global warming
closure 136
complementarity 75, 84
complex numbers 14
contingency 95
corpuscular theory 4, 44-46, 48-49,
   51-52, 104
Coulson, C. A. 75, 84
critical realism 142-143, 145, 147
Cudworth, R. 48, 62-63

Dalton, J. 10, 11, 69-71, 88
dark matter 106
Davy, H. 68
Dawkins, R. vii, 2, 110
de Broglie, L. 13-14, 22
de Lubac, H. 127-130
Democritus 9, 39, 44, 68, 88
Descartes, R. 44, 46, 47, 52, 58-59,
   62
determinism 49
Dirac, P. 3, 16, 22
dualism 4

Eddy, M. B. 5, 131ff
Einstein, A. 3, 12, 13, 15, 17-18, 20,
   26, 28, 94, 97, 105
Ellis, G. 146n.
emergentism 112
Empedocles 39, 40
entanglement 33, 34, 98, 104
Epicurus 39, 44
eschatology 125, 129, 147, 149

Everett, H. 22, 29-30, 34
evolution 52, 148

Faraday, M. 68
Frankland, E. 72

Gaia hypothesis 85, 96
Galileo 44, 51
Gassendi, P. 44
genes 2, 108
global warming 4, 76ff
Gnosticism 41-42
God
    action of 21, 46, 52, 63, 93-95,
        97, 119
    as creator 42, 60, 80, 90, 96,
        103, 113, 123
    as Trinity 50-51, 103, 111ff,
        122, 129-130, 149
    existence of 94
    of the gaps 51, 92, 95
Gould, S. J. 1
grace 5, 120, 123ff, 132
Gunton, C. 147

Heisenberg, W. 23, 26
    uncertainty principle of 24-25,
        32, 93, 104, 143
Heraclitus 39
Hesse, M. 67, 75, 83
Higgs boson 17, 63
Hobbes, T. 45-46
Hofmann, A. W. 71, 73
Hopkins, G. M. 1
Hugh of St Victor 50
humanism 126
Hume, D. 45, 148

indeterminism 4, 15, 21, 24, 27 33-
    34, 89, 90, 143-144
information 103, 106ff
Irenaeus 150n.

Kalām 25, 87ff
Kant, I. 148
Kekulé, F. A. 72

Kelvin, Lord, 67, 105
Kepler, J. 45
Kierkegaard, S. 75, 84
Kuhn, T. S. 138

Laplace, P.-S. 51, 52, 57, 63
Lavoisier, A. 69
Leibniz, G. 47-48
Locke, J. 46, 58-59
logos theology 5, 114-115, 121
Lovelock, J. 85, 96
Lucretius 39
Luther, M. 113

MacIntyre, A. 121-122
McGrath, A. 148
Manichaeism 42
materialism 2, 5, 48, 49, 57, 104,
    106, 129, 131
mathematics 3, 51
Maxwell, J. C. 10, 11, 12, 14
mechanical philosophy 45, 52
Mendeleev, D. I. 10
metaphor 4, 67, 68-71, 146-147
models 4, 67ff
Moltmann, J. 149
Monod, J. 110
Monti, A. 5, 141-142, 144, 145,
    146-147
More, H. 48, 62-63
Mutakallimūn 25, 87ff

narrative 2, 149
natural theology 1, 5, 51, 104, 113,
    121, 142, 147, 148ff
neo-scholasticism, 126-127
Newton, I. 2, 14, 15, 39, 47-48, 51-
    52, 59-60, 63, 68

occasionalism 46, 49, 59-60

Paley, W. 52
Pannenberg, W. 114
pantheism 112-113, 133
Paul, St 41, 50, 86, 139
Pauli exclusion principle, 16

photoelectric effect, 12-13
Planck, M. 3, 11-12
Plato ix, 39, 40, 42, 124
Polanyi, M. 145-146
Polkinghorne, J. 5, 93, 95-96, 97, 141n., 142ff
Priestley, J. 68
providence 40, 96
Puddefoot, J. 109, 120

quantum physics 3-4, 5, 9ff, 21ff, 52, 104, 142ff
qubits 107, 120
Qur'an 95, 97, 98

Rahner, K. 128-130
reductionism 53, 110
relativity theory 15, 16, 18, 104, 105
Russell, B. 104
Rutherford, J. 11

scholastic philosophy 43, 61
Schrödinger, E. 14-15, 22
  equation of 14-15, 22, 25, 26, 28, 30, 75
seminal principles 42-43, 49
Smith, J. M. 108, 110
Spinoza, B. 94
statistical mechanics 10
Steiner, G. 142, 147

Stoicism 40, 42, 49, 119-120
Swinburne, R. 148

tacit knowledge 145
Tait, P. G. 61
Taylor, J. 148
teleology 41, 129
Tennyson, A. 86
Tertullian 41
Thales 39
Thomson, J. J. 11
transubstantiation 46
Trinity, see God

ultraviolet catastrophe 10, 11
uncertainty principle, see Heisenberg

Vanstone, W. H. 147
von Neumann, J. 23, 27-28, 34

Ward, K. 2, 148
wave functions 14, 22, 27-29
wave-particle duality 13, 22, 75
Weinberg, S. 93
Wheeler, J. 26, 30, 34
Wigner, E. 3
witchcraft, 61-62
Wittgenstein, L. 5, 120-121
Wright, N. T. 81, 142, 146